New Technology-Based
Firms in the 1990s

New Technology-Based Firms in the 1990s

Volume IV

Edited by
WIM DURING
and
RAY OAKEY

Paul Chapman
Publishing Ltd

Selection and editorial material copyright © 1998 R. Oakey
Chapters © 1998 as credited

All rights reserved

Paul Chapman Publishing Ltd
144 Liverpool Road
London
N1 1LA

Apart from any fair dealing for the purposes of research or private study, or criticism or review, as permitted under the Copyright, Designs and Patents Act, 1988, this publication may be reproduced, stored or transmitted, in any form or by any means, only with the prior permission in writing of the publishers, or in the case of reprographic reproduction in accordance with the terms of licences issued by the Copyright Licensing Agency. Inquiries concerning reproduction outside those terms should be sent to the publishers at the abovementioned address.

British Library Cataloguing in Publication Data

New technology-based firms in the 1990s
  Vol. 4
  1. Technology innovations – Economic aspects – Congresses   2. New business enterprises – Congresses
  I. During, W.E.   II. Oakey, R. P. (Raymond P.)
  338'.064'09049

ISBN 1 85396 386 0

Typeset by Dorwyn Ltd, Rowlands Castle, Hants.
Printed and bound in the United Kingdom.

A B C D E F G H   5 4 3 2 1 0 9 8

Titles available in this series

New Technology-Based Firms in the 1990s Volume I
edited by Ray Oakey
1 85396 274 0 Hardback 1994 224pp

New Technology-Based Firms in the 1990s Volume II
edited by Ray Oakey
1 85396 343 7 Hardback 1996 208pp

New Technology-Based Firms in the 1990s Volume III
edited by Ray Oakey
1 85396 369 0 Hardback 1997 224pp

New Technology-Based Firms in the 1990s Volume IV
edited by Wim During and Ray Oakey
1 85396 386 0 Hardback 1998 208pp

New Technology-Based Firms in the 1990s Volume V
edited by Ray Oakey and Wim During
1 85396 387 9 Hardback 1998 288pp

Contents

List of Contributors ix

PART I INTRODUCTION

1. High-Technology Small Firms: Entrepreneurial Activity and the Co-operation Process — 1
   W. DURING AND R. P. OAKEY

PART II AN OVERVIEW OF NETWORK APPROACHES

2. The Theory and Practice of Innovative Networks — 10
   W. G. BIEMANS

3. Science in the Market Place: The Role of the Scientific Entrepreneur — 27
   K. DICKSON, A-M. COLES AND H. LAWTON SMITH

PART III CHARACTERISTICS OF HTSF-ENTREPRENEURS AND THEIR COMPANIES

4. Management Styles and Excellence: Different Ways to Business-Success in European SMEs — 38
   W. DURING AND M. KERKHOF

PART IV ENTREPRENEURIAL NETWORKING

5. Preconditions and Patterns of Entrepreneurial Networks in an Innovative Environment — 52
   K-H. SCHMIDT

6. Entrepreneurial Innovation Networks: Small Firms' Contribution to Collective Innovation Efforts — 62
   M. LARANJA

7. Support of Technology-Based SMEs: An Analysis of the Owner Manager's Attitude — 71
   M. KLOFSTEN AND A-S. MIKAELSSON

PART V TRUST IN FORMING AND OPERATING INNOVATIVE IORs

8. The Role and Means of Trust Creation in Partnership Formation between Small and Large Technology Firms: A Preliminary Study of how Small Firms Attempt to Create Trust in their Potential Partners 81
   K. BLOMQVIST

9. Strategic Alliances as an Analytical Perspective for Innovative SMEs 99
   M. MØNSTED

10. How Entrepreneurial Networks Can Succeed: Cases from the Region of Twente 112
    R. K. WOOLTHUIS, D. SCHIPPER AND M. STOR

11. Trust and Management: As Applied to Innovative Small Companies 125
    E. KRIEGER

PART VI CO-OPERATION BETWEEN SMALL AND LARGE FIRMS

12. Dynamic Complementarities with Large Advanced Companies: The Impact of their Absence upon New Technology-Based Firms 140
    M. FONTES

13. David and Goliath: To Compete or to Sell? 156
    G. M. P. SWANN

PART VII REGIONAL FACTORS AND THE HTSF

14. Knowledge Co-operation in the Dutch Pharmaceutical Industry: Do Regions Matter? 167
    M. VAN GEENHUIZEN

15. Inter-firm Links between Regionally Clustered High-Technology SMEs: A Comparison of Cambridge and Oxford Innovation Networks 181
    C. LAWSON, B. MOORE, D. KEEBLE, H. LAWTON SMITH AND F. WILKINSON

Contributors

W. BIEMANS, Faculty of Management and Organisation, University of Groningen, PO Box 800, 9700 AV Groningen, The Netherlands

K. BLOMQVIST, Kera Ltd, Snellmaninkatu 10, FIN-53100, Lappeenranta, Finland

A-M. COLES, Department of Management Studies, Brunel University, Uxbridge, Middlesex, UB8 3PH, UK

K. DICKSON, Department of Management Studies, Brunel University, Uxbridge, Middlesex, UB8 3PH, UK

W. DURING, University of Twente, School of Management Studies, Technology and Organisation, PO Box 217, 7500 AE Enschede, The Netherlands

M. FONTES, Instituto Nacional de Engenharia e Tecnologia Industrial, Estrada do Paço do Lumiar, 1699 Lisboa Codex, Portugal

M. VAN GEENHUIZEN, School of Systems Engineering, Policy Analysis and Management, Delft Technical University, PO Box 5015, 2600 GA Delft, The Netherlands

D. KEEBLE, ESRC Centre for Business Research, University of Cambridge, Sidgwick Avenue, Cambridge, CB3 9DE, UK

M. KERKHOF, Ministry of Transportation, Transport Research Centre, Freight Transport Division, PO Box 1031, 3000 BA Rotterdam, The Netherlands

M. KLOFSTEN, Linköping University, Centre for Innovation and Entrepreneurship (CIE), HusETT, S-581 83 Linköping, Sweden

E. KREIGER, HEC School of Management, 1 Rue de la Liberation, 78351 Jouy-en-Joasa Cedex, France

M. LARANJA, AITEC-Enterprise Innovation Centre, Av. Duque d'Ávila 23 lo Dto 1000, Lisboa, Portugal

C. LAWSON, ESRC Centre for Business Research, University of Cambridge, Sidgwick Avenue, Cambridge, CB3 9DE, UK

A-S. MIKAELSSON, Linköping University, Centre for Innovation and Entrepreneurship (CIE), HusETT, S-581 83 Linköping, Sweden

M. MØNSTED, Department of Management Politics and Philosophy, Copenhagen Business School, Blaagaardsgade 23 B, DK-2200 Copenhagen, Denmark

B. MOORE, ESRC Centre for Business Research, University of Cambridge, Sidgwick Avenue, Cambridge, CB3 9DE, UK

R.P. OAKEY, Manchester Business School, Booth Street West, Manchester, M15 6PB, UK

D. SCHIPPER, Demcon Twente b.v., Westermaatsweg 11, 7556BW, Hengelo, The Netherlands

K-H. SCHMIDT, University of Paderborn, Department of Economics, 33100 Paderborn, Germany

H. LAWTON SMITH, Centre for Local Economic Development, Coventry University, Priory St, Coventry, CV1 5FB, UK

M. STOR, Innovation Centre Network Netherlands, PO Box 5503, 7500GM, Enschede, The Netherlands

G. M. P. SWANN, Manchester Business School, Booth Street West, Manchester, M15 6PB, UK

F. WILKINSON, ESRC Centre for Business Research, University of Cambridge, Sidgwick Avenue, Cambridge, CB3 9DE, UK

R.K. WOOLTHUIS, University of Twente, Postbus 217, 7500 AE Enschede, The Netherlands

PART I Introduction

CHAPTER 1

High-Technology Small Firms: Entrepreneurial Activity and the Co-operation Process

WIM DURING AND RAY OAKEY

INTRODUCTION

This is the fourth volume in what is now a well established series based on the best papers of the annual High Technology Small Firm (HTSF) conference. After three very successful meetings at Manchester Business School, a fourth conference was held at the University of Twente in The Netherlands during September 1996. This change of venue bears witness to the extending international character of the conference.

It is the intention of Ray Oakey at Manchester Business School (MBS) and Wim During at The University of Twente (UT) to continue to co-operate on a regular basis in organising this forum for presenting and discussing research findings and practitioner experiences pertinent to the development of HTSFs. We see our co-operation as a small metaphorical example of two important issues, critical to the success of HTSFs: namely co-operation and internationalisation.

This fourth conference was organised around the theme of 'Entrepreneurial activity and the co-operation process'. In the past volumes of this series, a growing interest has been evident in alliances, networking and partnering. For example, in Volume III, this general topic area comprised a section of six papers.

Such growing interest in the co-operative actions of HTSFs within this series reflects recent trends in research into entrepreneurship and small business management in Europe. In a recent review of research on entrepreneurship and small business management, co-operation emerged as one of the four most researched themes (Landstrom et al., 1997). However, despite the growing body of research in this field, much remains unclear. But this is a natural outcome of the way researchers are dealing with the issue, since co-operation is studied at several organisational levels, and from different theoretical view points.

It is clear however that, for networking and partnership concept to be relevant to HTSFs, the specific character of this type of firm must be borne in mind. For small firms in general, the dominant role that the entrepreneur plays *within* the firm, and in promoting co-operation *between* firms, is a key determinant of success. However, in the case of HTSFs, the entrepreneur often plays a particularly crucial role in determining both the basic strategy of the enterprise, together with developments in the technical field. Therefore, varying qualities of technology driven entrepreneurs provide the basic opportunities and the constraints which respectively aid or restrict co-operation between HTSFs. This point is not always understood by external stakeholders dealing with HTSFs.

The developments causing the growing importance of, and thus interest in, co-operation and interorganisational relationships between firms have been much discussed. In this context the crucial role of co-operation has, of course, been cited by research into the preconditions necessary for successful innovation. The importance of external communication, and the establishment of links with outside organisations, has been widely noted for small firms in general, and HTSFs in particular. The basic attraction of today's interest in co-operation is that this interaction will have positive synergistic effects, ramifications that are a necessary condition for success in today's competitive markets.

The benefits of co-operative innovation processes, classed by different authors, broadly fall into four categories (see for instance Chesnais, 1988; Jaquemin, 1991; Hakanson, 1993; Bidault and Cummings, 1994). First, synergies in technological collaborations may lead to shorter development processes, promote common standards, and get access to knowledge in new specialist fields. Second, synergies in management may be derived from co-operative training and making use of each other's 'best practices'. Third, shared resources, reduced risk, and the attainment of critical mass, are other economic benefits associated with co-operation. Finally, at a strategic level, co-operation may facilitate access to new markets, reduce competition, and enhance the flexibility of the partners.

However, although the potential benefits are many, the above and other authors signal that co-operation, desirable in principle, often leads to problems in practice. Problems with intellectual property ownership occur when unauthorised technology transfers take place between collaborating firms, or when one of the partners has insufficient technological knowledge to create a balanced collaboration. Managing the co-operation process often proves difficult in the absence of a formal agreement of roles performed by partners, which can cause friction, for instance, when the persons managing the co-operation change over time. Differences in culture or opportunistic behaviour are further potential sources of conflict. Conflicts may also arise over questions of sharing the intellectual property ownership of collaboration results, while sometimes, time consuming arguments over ownership issues erode the expected benefits. Strategically, entrepreneurs engaged in collaborations often have difficulty in accepting the loss of independence regarding decision making, or in allowing a partner access to their 'own' market.

Although the outcome of co-operations may be positive or negative, actively innovating companies seem to have no choice but to co-operate to a greater or lesser extent, in the best way possible. In the 1996 edition of an annual 'Technology Lecture' organised by the Ministry of Economic Affairs in The Netherlands, Mr. Augustine, president of Lockheed Martin, summarised the conditions in which HTSFs do business as follows:

> What, then, is the answer to the dilemma of managing innovation in a world that consists of impatient investors, costly projects, and rapidly advancing technology? The answer of course is that we need innovative approaches to innovation.

Augustine then goes on to describe three techniques in use within Lockheed Martin that also might be used by co-operating HTSFs. The first concept is that of the virtual company which, in essence, is the bringing together of all those resources needed for a particular project. The second concept concerns sharing critical resources with other corporations. An example of this phenomenon is given in which each of two co-operating companies may use the outcome of joint research, provided that they do not compete in each other's marketplace. The third concept involves creating and using specialised coalitions between the company and selected academic institutions. For Augustine, 'reaching out' is a key part of a HTSF's life:

'No matter what your size, you have to look for new ideas, new approaches'.

All three approaches mentioned by Augustine suggest the importance that must be ascribed to co-operation within or between companies. Given this perceived importance of co-operation and the forming of networks, governments and support agencies who see a role for themselves in stimulating innovation should promote, or even demand, co-operation between firms when promoting publicly funded projects. This is evident on both national and European scales, which raises the question of the appropriateness and timing of government policy delivery in this context.

As noted in Volumes II and III, there is evidence of a 'gap' in policy making between what we think should happen in theory and what happens in practice. In this respect, before we can develop effective policies, it is important to create better insights into the phenomenon of co-operative innovation and of the venture creation process. For HTSF managements, pertinent questions might be: what forms of co-operations are effective, and which are not, and under what circumstances? Which types of collaborations can be improved by the selection of the right partners, and what projects might be aided by directing and managing the development of the co-operation process?

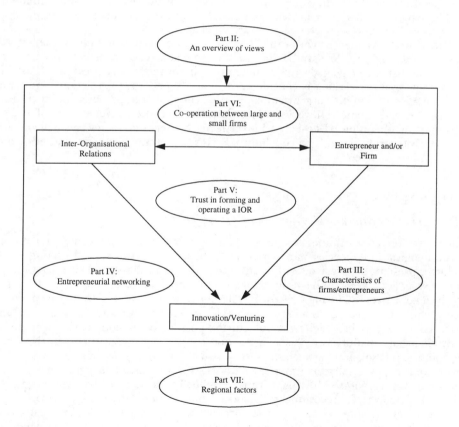

Figure 1.1
Major themes and interrelations of the innovation and venture creation process typical of HTSFs and their entrepreneurs

Moreover, what are the important conditioning parameters for the above phenomena? In addition to these questions, it is important for supporting development organisations to have knowledge of new and emerging effective support strategies. For governments, an additional question is whether co-operation in general, or in specific instances, is something that really 'deserves' policy support.

The entrepreneurial process of creating, developing and exploiting technology-based market opportunities is a core activity, in which all HTSFs are engaged. Innovation, venture creation and business creation are the underlying processes involved. The execution of these processes during co-operations with other firms (or institutions) will follow the 'laws' of Inter Organisational Relationships (IORs) with their dynamics a balance between different interests, or a tension between power domination and a reliance on trust. The specific backgrounds of the HTSFs, and their entrepreneurs, will influence the way in which they operate these processes. Personal background, the development stage of the firm, and the specific technical expertise of the entrepreneur, will all play roles. The result will be that IORs take many different forms. Figure 1.1 displays the characteristics of the innovation and venture creation process, typical of HTSFs and their entrepreneurs. This figure relates major aspects of this innovation and collaboration process within HTSFs to the major themes under which papers are grouped in this volume, together with an indication of the relevant chapter numbers.

The papers in this book present new insights into the dynamics and development processes of specific co-operative innovation collaborations.

Part II presents an overview of the literature on co-operation and innovation over the past two decades, while Part III focuses on the characteristics of entrepreneurs and their companies, engaged in co-operative innovation activities. In Part IV, entrepreneurial networking is considered. In Parts V and VI, the relationship between the characteristics of interorganisational relations (and of entrepreneurs and their firms) are the central theme. However, while Part V covers the development of trust in forming and operating IORs, Part VI investigates co-operation between small and large firms as a central theme. The concluding Part VII presents pertinent findings from a regional perspective.

THE PAPERS

An overview of network approaches (Part II)

Literature on co-operation and networking is often rather confusing. To put the terminologies and theories into perspective, Biemans, in the second section of this volume, presents an overview of literature regarding innovation and co-operation over the past twenty years. The terminologies and forms of co-operation, as they have evolved concurrently during the past two decades, are discussed. During this period we see a shifting focus. At first, relationships between two organisations are the main concern, a context in which a manufacturing perspective is dominant. Customer involvement and manufacturer-user interaction are the key themes here. In the second period, the relationship between more than two organisations becomes important. The main concern, especially for American authors, is now strategic management, strategic alliances and networks. Most papers in the following sections of this volume adopt a network perspective. According to Biemans, the relationships between two or more parts of an organisation become more central in the next stage with virtual organisations and co-opetition providing the new key words. The perspective from which co-operation is studied is then broadened to encompass many scientific disciplines, including economics, sociology and organisation theory.

Based on a previous overview of mainstream findings regarding co-operation and innovation, Biemans concludes by offering some guidelines for managers engaged in co-operation activities.

Characteristics of HTSF-entrepreneurs and their companies (Part III)

A relatively recent development in research on entrepreneurial activity is the concept of entrepreneurial orientation. It acknowledges the differences in ambition and action-orientation of entrepreneurs regarding autonomy, innovativeness, risk taking, pro-activeness and competitive aggressiveness (Lumpkin and Dess, 1996). It is their contention that, during interaction, entrepreneurial orientation, organisational factors and environmental factors determine the performance of a firm.

The papers in this section present a further specification of entrepreneurial orientation in the context of HTSFs in the pursuit of innovative ventures in a co-operative context.

Based on previous literature, and 6 UK based case studies, Dickson et al., in the first paper of this section, specify the role of the scientific entrepreneur in commercialising the results of scientific advances. Due to differences in personal orientations, and differences between academic and business cultures, science oriented entrepreneurs seem to favour one of three different routes to entrepreneurship:

1. The academic entrepreneur engages in entrepreneurial endeavours, but only as adjuncts to their academic work.
2. Entrepreneurial scientists keep their focus on scientific values and work, but pursue these goals through a business that draws on their expertise.
3. The scientific entrepreneur displays both scientific and business acumen, and pursues new business opportunities utilising their scientific intelligence.

During and Kerkhof present more in-depth information on entrepreneurial approaches, and their relationship to different levels of innovativeness and types of learning styles. Their study is based on data from 227 manufacturing firms, derived from a European survey of successful small firms. Regarding their strategies, the more innovative firms in this sample (more active in terms of self-generated new product development) were more aggressive in entering new markets than the other firms. However, the intensity of engaging in IORs on strategic and operational levels was the same in all survey firms. The strategic content was quite different, however, between the more and less innovative firms. The groups follow different strategies in the fields of co-operative product design and operational matters regarding joint purchasing or distribution. They also differed with respect to who takes the lead role in 'contracting out' and/or 'contracting in' joint production activities. The fact that the innovative companies use project management as a regular management tool was seen to have a positive effect on their ability to successfully act in co-operative development activities.

Entrepreneurial Networking (Part IV)

Entrepreneurial networking is a means by which HTSFs develop new activities or ventures in co-operation with other firms. In organising and operating a network, it is necessary to pay particular attention to general problems that might inhibit the success of the innovation processes. In particular, problems arise from uncertainty about outcomes, iterative learning processes, and the availability of highly educated personnel. A specific characteristic associated with co-operation between HTSFs is the reality that the entrepreneurs will, themselves, operate as actors in the network and will often be involved in the execution of project activities.

By providing a view of the entrepreneurial process, the paper of Schmidt concentrates on the way co-operating suppliers operate in the face of technical changes. His analysis is based on data from Germany, Switzerland, Japan and South Africa. Four types of entrepreneurial network are apparent, not all of them being formal networks. The kind of network a firm participates in is, to a large extent, dependent on the innovative orientation of the management. The different focuses of these networks in the fields of knowledge creation and transfer, production, marketing/sales and finance, are also illustrated. While networks in Japan tend to be more vertically oriented, in the other countries of this research they tend to be more horizontal. In Germany and Switzerland, Schmidt particularly notes that rigidities in the dual system of vocational education pose problems in the fields of accelerated technical change.

In the second paper of this section, Laranga presents an example of the role of NTBFs in disseminating new technology into diverse areas of application. The study focuses on the entrepreneur's function in identifying opportunities to translate upstream technology, developed elsewhere, into localised products.

The study makes it clear that the integration of R&D, marketing and sales with manufacturing is normally a condition required for successful innovation. However, for a less advanced country, such as Portugal, these functions are realised through a complex network of formal and informal interactions, involving small technology based firms, local universities, and the local subsidiaries of multinational companies. By establishing personal linkages to universities and large users in retailing and banking, the entrepreneurs in this research, through identification, adaptation, service and training, have made new technology available for users in the local area. The acumen of the high technology entrepreneur in designing a network that meets these demands is an important parameter for success.

Apart from direct experience, another way to develop the necessary skills is to participate in the development programmes of support organisations. Such assistance can play an important role in developing the insights and abilities required to operate entrepreneurial networks. In the concluding paper of this section, Kloften and Mikaelsson report the experiences and attitudes of entrepreneurs in the Linköping area of Sweden. Their findings are in line with other studies into the use that entrepreneurs make of support organisations. Only entrepreneurs with a reasonable level of management sophistication (i.e. a resource base, internationalisation experience) use the services of formalised interactive programmes. Entrepreneurs view the improvement of networking and communication between firms as the most important role of support organisations.

Trust in forming and operating innovative IORs (Part V)

Trust is often recognised as an important condition for successful co-operation between different organisations. Yet the nature of trust, involving how the forming and sustaining of trust influences the process of co-operation, how the working of trust can be researched in a co-operative setting, remains elusive. For HTSFs, the issue of trust is felt to be important in developing a position in the marketplace. As a consequence, a line of research is developing into the role of trust in the co-operative activities of HTSFs. The second and third volumes of this series have each presented one paper with a focus on trust. This section of the fourth volume comprises four papers, all of which address the theme of trust. Furthermore all four papers present results from case studies into the role of trust in different forms of IORs, in which HTSFs are involved.

The first paper, by Blomqvist, has a strong conceptual orientation, focusing on the nature, the role and the operationalisation of trust. One basic viewpoint is that trust is essentially something perceived by actors, under the influence of experiences, social learning and cultural backgrounds. During a co-operation, trust may become the outcome of a process evolved during the partnership. The essence of co-operation and trust can be captured by the following statements: the building of trust is difficult to imitate, it is slow to grow, it is easy to break, and is difficult to heal when broken. In technical partnerships, with their inherent uncertainties, the role of trust is expected to be especially important.

Trust is defined here by Blomqvist as 'the actor's expectations of the other party's competence and goodwill'. Both competence and goodwill are operationalised and used in an exploratory study into the way managers of large firms, and entrepreneurs of small firms, go about creating trust.

Co-operation in the development of new products is part of a strategic process. For the alliance of small technology-based firms, however, Monsted argues that the existing literature on strategic alliances is of little use, because of its focus on large firms. In a study of thirteen HTSFs in the biotechnology, pharmacy, and computer software sectors, the paper develops a small firm perspective to this field. The subject under study was how some Technology Based Firms (TBF) prove legitimacy and communicate their potential to larger potential collaborator enterprises in the early stages of their development. For small TBFs, credibility is low during their early life. They seek out possibilities to co-operate as one way to gain credibility, and thus get access to trust-based networks. One of the products of this behaviour is, for the young TBF, a strategy establishing what to accomplish through co-operation, after they have begun to co-operate. A strategic alliance is thus not designed, but is the result of a developmental process. During this evolution trust plays an important role. The issue of trust is also highlighted from a developmental viewpoint. Especially in its early stages, a NTBF must rely on trust as a means of earning its place in a network with more powerful larger partners.

In the next article, Woolthuis et al. give further insights into the process of developing an entrepreneurial network, and the way in which trust plays a role in this process. Co-operation is not an activity that comes naturally to most companies. For entrepreneurial networks, with their focus on new business development and inherently complex projects, co-operation is achieved very much through a learning process. Within networking activities, entrepreneurs must learn to interpret verbal and non-verbal communication. Trust is important in overcoming early misunderstandings, while developing understanding is a precondition for trust in subsequent activities.

Innovation Centres have diverse experiences in assisting entrepreneurs in setting up and developing innovative networks. These experiences are highlighted using a specific case of nine SMEs and three government institutes. The aim of the partners in the network is to create new business in the field of medical support products, to gain more experience in the field, to gain experience in co-operation, and for the younger companies to become known in the marketplace. The paper concludes by drawing on the experiences of the Innovation Centres to arrive at the 'ten commandments for co-operation'.

The final paper in this section presents yet another way to create trust in the context of innovative new companies. Krieger presents trust as a positive anticipation associated with an assumed risk. In line with the other papers in this section, trust is viewed from a process perspective. Trust building is a process that develops in three main directions: experience, perceived proximity and guarantees. Possible facilitators of trust are thus a positive reputation, a common culture, shared values, and the existence of formal validation procedures. Trust building can be managed in a positive way by facilitating information flows, while maintaining confidentiality by increasing the frequency of interactions between partners to reduce real or perceived risk, and by securing each interaction through the use of a third party as a real (or symbolic) guarantor. Finally Krieger draws our attention to the pivotal role of the entrepreneur in his or her inclination (or disinclination) to trust a partner.

Co-operation between small and large firms (Part VI)

One of the recurrent themes in HTSF research is their co-operation with large firms or institutions involving different complementarities. In this section Fontes takes an interesting angle, which begins with the observation that there is an absence of large advanced firms in the Portuguese economy. Therefore, due to their absence, the process of 'spin-off' or 'incubation' is less prominent in this national case. The absence of their larger firms also denies NTBFs a 'natural' access to the sponsoring and main customer roles of the incubator company. The paper looks into the alternative practices developed by 28 NTBFs mainly in the field of information technology. These NTBFs concentrate on looking for other firms, for whom the specific NTBF know-how of others is a welcome source of advanced knowledge, ideas and products. The strategy that appears most successful, under these circumstances, is to customise products developed elsewhere, for use under local conditions. Most often, the partners of NTBFs were large service organisations, aware of the possibilities of new technology, but without the necessary technical knowledge and skills. Taking part in the development of advanced new products was notably absent in the Portuguese case.

When advanced large companies are available for collaboration with HTSFs, another strategic choice poses itself, involving whether to co-operate or to compete 'head on'. In the last paper of this section, Swann addresses this problem of introducing new technology into a market in an analytical way. As a tool to be used by HTSF entrepreneurs, he systemises the pertinent questions, decisions and possible outcomes. It is argued that, when the industry standard is not yet settled, a joint venture would, in many instances, expand the overall market. Otherwise, a niche strategy would be more likely to be a successful strategy for the small firm.

Regional factors and the HTSF (Part VII)

While the previous sections were presented as empirical and theoretical findings about co-operation of HTSFs, primarily from the perspective of the individual firm or entrepreneur, the two concluding papers of this volume take a regional perspective. With her study of co-operative strategies in the Dutch pharmaceutical industry, Geenhuizen presents the theoretical concept of the innovative milieu. Based on the empirical facts of the industry, the argument of her paper is mainly of a conceptual nature. Starting from the process of pharmaceutical product development, it is argued that a local/regional orientation is preferred for co-operative activities. Three stages are proposed for this process including discovery and synthesis of potential products, pre-clinical testing, and technical processing. To improve co-operation on a regional basis, it is proposed that facilities should be developed for 'spin-off' companies in the biotechnology field, but also to match educational programmes in universities to specific subjects of interest for new product development.

In the final paper Lawson et al. present a study of the inter-firm links of regionally clustered firms. This study is also relevant to the innovative milieu literature noted above. The sample for this study consisted of 100 firms (50 in and around Cambridge, and 50 in the Oxford area). The study shows that close inter-firm links within the region are predominantly of the supplier/subcontractor and service provider type. Outside the region, customer links provide the most frequent contacts. Several differences between Oxford and Cambridge HTSFs are apparent. For instance, in managing the risks associated with inter-firm links, the Oxford firms relied more on personal relationships and experience of fair trading, whereas the Cambridge firms made more use of legal documents and compliance to contracts. The propensity to form close inter-firm links between HTSFs was greater in the Cambridge area, maybe due to the larger number of HTSFs in this region. For both regions, evidence of innovative milieu factors was present, one example being the number of 'spin-off' companies from larger firms, where the 'spin-off' firms were mostly located in the same region. Generally, for both regions, the importance of local sources for ideas and the development of innovative activities were confirmed, and in line with previous research findings. On average, links with local universities were less frequent than might have been desired by policy makers. In the Cambridge area, university links appeared less important than in Oxford. This may reflect the fact that there were a larger proportion of companies begun more than five years ago in the Cambridge case.

The last finding points to the development perspective which is an important issue in the interorganisational relationships of HTSFs. A theme that runs through many of the papers in this volume concerns the developmental process of IORs. At the same time, in most of the studies presented here, IORs are formed between a mixture of new, or recently founded firms, and more mature firms. As a consequence, not only the IOR develops over time, but within this context, firms and their entrepreneurs evolve through different stages of their life cycles. In addition, the IORs are embedded in regional developments, which partly determine the feasibility of different kinds of alliances and ventures. Building on the research presented in this volume, further investigation is needed into the development and dynamics of the co-operation process within and between HTSFs. Understanding this process, as it develops, is a necessary but not sufficient condition. Apart from the process itself, the outcome in terms of new venture success must also be part of this research agenda. The results of such research will help both firm managements, as well as policy departments of government (at various levels), to design and execute effective strategies and programmes to promote greater co-operation.

REFERENCES

Augustine, N.R. (1996) The innovative enterprise in: organising innovation, in search of adaptability, technology lecture, ministry of economic affairs, The Hague, 12 December.
Bidault, F. and Cummings, T. (1994) Innovation through alliances: expectations and limitations, *R&D Management*, 24, 1.
Chesnais, F. (1988) Technical co-operation agreements between firms, *STI Review*, 4, December.
Hakanson, L. (1993) Managing co-operative research and development: partner selection and contract design, *R&D Management*, 23, 4.
Jaquemin, A. (1991) Co-operation in Research and Development and European Competition Policy in De Wolf, P. (ed.) *Competition in Europe*, Kluwer Academic.
Landstrom, H., Frank, H. and Veciana, J.M. (eds.) (1997) *Entrepreneurship and Small Business Research in Europe*, Avebury.
Lumpkin, G.T. and Dess, G.G. (1996) Clarifying the entrepreneurial orientation construct and linking it to performance, *The Academy of Management Review*, 21, 1.

PART II An Overview of Network Approaches

CHAPTER 2

The Theory and Practice of Innovative Networks

WIM G. BIEMANS

INTRODUCTION

The last decade has been typified by a business environment that seems obsessed with the subjects of innovation and networks. Both academic journals and the business press abound with stories about the creation of new products and services through the collaboration of numerous partners. The automobile industry set the trend by making close cooperation with suppliers the standard way of doing business, while airlines join forces with foreign partners in order to remain airborne in an increasingly competitive market. Moreover, manufacturers of computers team up with producers of consumer electronics and traditional suppliers of entertainment, to create exciting new multi media products.

This paper traces these new developments, as well as the relationship between the various concepts and buzzwords introduced to describe and explain this new management reality. In addition, it discusses the major issues facing firms that are trying to adapt to this new organizational environment by joining in the game of cooperation, with special emphasis on the plight of small firms. Since many firms are still struggling to make cooperation work, the paper closes with a challenging agenda for future research in this area.

TERMINOLOGY AND CONFUSION

Drastic changes in the business environment, such as intensified competition, a growing complexity of products and production processes, shortened product life cycles and increasingly sophisticated customer demands, have resulted in a fascination with innovation and networks. This growing attention to innovation, cooperation and networks has been accompanied by the introduction of a large number of theoretical concepts, buzzwords and management fads. Numerous management thinkers have attempted to make their mark by introducing their own terms. Since many of these newly coined catchphrases refer to the same (or at the very least) quite similar phenomena, their proliferation has resulted in chaos and confusion. This section will seek to reduce this confusion by grouping the various terms into 3 categories and explaining how they are related. The 3 categories comprise:

1. Relationships between 2 organizations

2. Relationships between more than 2 organizations
3. Relationships between two or more parts of a single organization

Relationships between two organizations

Most of the literature on cooperation concerns the relationships that exist between 2 organizations. It all started with studies of *buyer-seller relationships* in industrial markets, which emphasized the mutual dependency of both parties. The characteristic of mutual dependency has caused some writers to talk about *co-destiny relationships*. In the marketing field, this has resulted in the practices of *account management* and *relationship marketing*. Looking at the relationship upstream (i.e. with one's suppliers) led to such concepts as *co-makership*, *co-design*, *just-in-time purchasing*, *supplier development* and *supply chain management*.

Others have taken these concepts out of the industrial marketing context and applied them to other business situations. Thus, the more general idea of *partnerships* and *strategic alliances* was born. The central thought behind these kind of relationships is that 2 organizations create value by joining forces (which has led some researchers to introduce the term *value-adding partnership*).

Relationships between more than two organizations

Since most firms do not restrict themselves only to collaborating with one other organization, the concept of a dyadic relationship was soon expanded to describe the constellation of relationships that a firm may pursue with various other organizations (such as key suppliers, major customers, competitors and universities). In its most general form, this is referred to as a *network*, *cluster* or *consortium*, with other management writers insisting on using more sophisticated names, such as *business ecosystem*, to describe the same phenomenon. The concept of networks is accompanied by a number of corresponding terms to describe the new organization. Some refer to it simply as the *network organization*, while others prefer more imaginative terms, such as *extended enterprise*, *hybrid organization*, *modular organization*, or even *Rolodex organization*, *cyberspace corporation* or *shamrock organization*. To emphasize the disappearance of once-clear organizational boundaries, management writers refer to the *boundaryless organization* or the popular *virtual organization*.

Relationships between two or more parts of an organization

The final group of terms refers to the internal relationships between various parts of an organization. The most visible form is the *Marketing-R&D interface*, which plays a major role in successful product development. The study of product development within Japanese firms has led to the development and adoption of concepts such as *concurrent engineering*, *parallel product development*, *rugby approach*, *fast track product development*, and the, by now familiar, *cross-functional team* or *multi-functional teams* approaches.

These 3 categories of concepts are obviously closely related. The network concept is a straightforward conceptual extension of the dyadic relationship. The connection between external and internal relationships is of a more complex nature. Although external and internal cooperations are typically treated separately, several authors believe them to be closely related. For example, based on an empirical investigation into product development within the Dutch medical equipment industry, Biemans (1991) noted that "the functioning of each of the internal networks directly influences the efficiency and efficacy of the external network". Dan Ciampa, CEO of the Rath &

Strong Consulting put it as follows: "Unless an organization knows how to foster collaborative relationships internally, it won't be good at making such relationships outside" (Magnet, 1994).

The complex interaction between internal and external relationship is acknowledged by several firms which strive to devise programs to integrate both kinds of cooperation into one successful approach. For instance, in developing the innovative Boeing 777, Boeing successfully included both key suppliers and major airlines in their numerous build-design teams (Sabbagh, 1996). This approach blurs the distinction between internal and external members of a product development team, and recently resulted in the introduction of yet another term: *virtual teams* (Lipnack and Stamps, 1997).

FROM CUSTOMER INVOLVEMENT TO CO-OPETITION

Having explored the confusing terminology concerning cooperation, this section traces its evolutionary development in organizational theory from customer involvement to co-opetition. This evolution consists of 6 subsequent stages:

1. Customer involvement in initiating product development
2. Buyer–seller relationships
3. Strategic alliances
4. Networks
5. Virtual organizations
6. Co-opetition

These 6 stages will be described briefly.

Stage 1: Customer involvement in initiating product development

Traditionally, the management of product development has typically been viewed from the manufacturer's perspective. The process of new product development is usually divided into a series of stages, and various empirical studies were designed to formulate specific recommendations for the manufacturer to improve its management of the new product development process and increase the likelihood of new product success. According to this perspective, the manufacturer is the only party involved, who controls the process and influences the environment. Famous examples of such studies are the Booz-Allen & Hamilton investigations of product development (1968, 1982).

In 1976 and 1977 von Hippel published the results of 2 empirical investigations which demonstrated that, in some industries, it is the user, rather than the manufacturer, who initiates the development of a new product. Based on his studies of mainly small firms in the scientific instruments, semiconductor and electronic subassembly industries, von Hippel (1978) formulated the *customer-active paradigm* (CAP) as the counterpart of the *manufacturer-active paradigm* (MAP). The MAP describes the more traditional situation mentioned above, where the manufacturer employs traditional methods of market research to investigate user demands and uses the information thus gained to guide new product development. In other words, the user is only passively involved. With the CAP, on the other hand, it is the user who has an idea for a new product and who initiates the development process by approaching a manufacturer with the proposal to transform the idea into a new product and launch it in the market. Sometimes, a user may even go further by developing a crude prototype, and testing it in practice. This led von Hippel to suggest the focussing on so-called *lead users* (von Hippel, 1986; Herstatt and von Hippel, 1992).

Stage 2: Buyer-seller relationships

Von Hippel's findings inspired a large number of similar studies by researchers who demonstrated user involvement in product development processes in various industries (Parkinson, 1982; Foxall and Tierney, 1984; Voss, 1985; Shaw, 1986; vanden Abeele and Christiaens, 1987; Biemans, 1991; Ciccantelli and Magidson, 1993). However, these researchers did not, *a priori*, limit the user's involvement to the idea generation stage of the product development process, and consequently the focus gradually shifted from characterizing the product development process as being either manufacturer-active or customer-active to determining the role of users during the whole process of product development.

Apart from these product development studies, buyer-seller relationships have gained dominance within the field of industrial marketing. Originally, the field of industrial marketing was derived from consumer marketing as industrial companies applied the familiar concepts and tools of consumer marketing to their industrial setting. In the second half of the 1970s, a number of European academics joined forces in a research group called the Industrial Marketing and Purchasing (IMP) Project Group and started to investigate buyer-seller relationships in industrial markets. Rather surprisingly, they found quite stable, long-term relationships between buyers and sellers, and developed a relationship perspective on industrial marketing, as opposed to the traditional transaction perspective (Håkansson, 1982; Hallén and Johanson, 1989). By including the exchange of social values, the IMP interaction approach is conceptually richer than Williamson's transaction cost approach (Williamson, 1975).

Later, the concept of buyer-seller relationships has been used to investigate supplier-involvement in product development (Asmus and Griffin, 1993; Håkansson and Eriksson, 1993), and to develop partnerships with suppliers in general (Sheth and Sharma, 1997). Management scholars focused on buyer-seller relationships to study the workings of the (Japanese) automobile industry and to explain the success of its major players (Dyer and Ouchi, 1993; Richter and Wakuta, 1993; Dyer, 1996). Nowadays, the practices of Japanese car manufacturers have been emulated by a number of Western companies, as witnessed by the current success of Chrysler and its outstanding supplier relationships.

Stage 3: Strategic alliances

Around the mid-1980s, an increasing number of American authors emphasized the growing prevalence and importance of strategic alliances (see e.g. James, 1985). This work built upon the concept of buyer-seller relationships and linked it more explicitly with corporate strategy. Strategic alliances may serve a broad range of strategic purposes, such as developing new products, exploiting technology, swapping products and gaining access to foreign markets. Because of their growing importance and broad applicability, strategic alliances have been the subject of a veritable flood of publications, both in the general business press and in scientific journals. Academic contributions to the subject of strategic alliances can be divided into 2 categories.

1. *Descriptive studies* that try to explain strategic alliances and formulate conceptual tools to study them. For instance, a Dutch group of scientists developed a general framework for studying strategic alliances (Huyzer et al., 1990). This framework was subsequently used to analyse the use of strategic alliances by a number of firms in the Netherlands (Huyzer et al., 1991). Others explain the functioning of strategic alliances by linking them to the concept of embedded knowledge (Badaracco, 1991), or by focussing on specific kinds of alliances, such as the ones between large and small firms (Botkin and Matthews, 1992).
2. *Normative studies* try to identify the factors distinguishing success from failure with

regard to strategic alliances. Generally, the practical guidelines for successfully developing and managing strategic alliances are formulated from research based on limited anecdotal evidence (Devlin and Bleackley, 1988; Sonnenberg, 1992; Lynch, 1993), while other investigations have followed a more rigorous approach (Niederkofler, 1991; Lorange and Roos, 1991, 1992; Bronder and Pritzl, 1992; Slowinski et al., 1993; Millson et al., 1996; Spekman et al., 1996). Most researchers describe the management of strategic alliances according to a limited number of growth stages.

Stage 4: Networks

The 1980s have witnessed a surging interest in the practice of various organizations joining forces to attain their own corporate goals. This behaviour is generally referred to as a network, and has been studied by scholars in a number of different disciplines such as Sociology, Economics, Industrial Marketing, Purchasing and Adoption and Diffusion Theory (Biemans, 1996).

Sociologists have used networking concepts to analyze structures such as patterns of informal communication within organizations and friendship networks. This has mostly resulted in rather formalized and quantitative approaches (Burt, 1982; Burt and Minor, 1984). In economics, concepts such as clusters (Porter, 1990) and the Japanese *keiretsu* (*Harvard Business Review*, 1990; Cutts, 1992; Burt and Doyle, 1993; Miyashita and Russell, 1994) are used to analyze the success of industries, regions and even nations. Recently, Moore (1996) called such networks *business ecosystems* and explains their functioning by using analogies from biology.

In the field of industrial marketing, the Swedish branch of the IMP Group transformed the interaction perspective into a network approach, with the focus shifting from buyer-seller relationships to the pattern of relationships which surrounds a firm (Håkansson, 1987; Anderson et al., 1994; Ford, 1997). Since the mid-1980s, an increasing number of academics have applied the concepts developed by the IMP Group to various areas of industrial marketing (see e.g. Biemans, 1992). In the related area of purchasing, drastic reductions in the number of suppliers, and closer ties with those remaining, have transformed purchasing into supply chain management (e.g. Saunders, 1994; Gattorna and Walters, 1996). New approaches are being developed to manage a clearly structured network of suppliers and subcontractors, with an emphasis on trust, cooperation and co-makership (Olsen and Ellram, 1997). Leenders and Blenkhorn (1988) introduced the term *reverse marketing*, which describes a proactive approach to purchasing by identifying potential suppliers and offering suitable supplier partners a proposal for long term collaboration (see also Biemans and Brand, 1995).

In adoption and diffusion theory, network concepts have been employed for 50 years, with famous examples being the study of the 1940 US presidential election (Lazarsfeld et al., 1948) and the investigation of the diffusion of a new pharmaceutical drug (Coleman et al., 1966). Rogers (1995) integrated the findings of many studies and distinguished between 5 categories of adopters, who subsequently adopt a new product or idea over time. The complexity of social systems and diffusion patterns have caused some researchers to apply network analysis to this area (Woodside and Wilson, 1994; Biemans and Setz, 1995).

Stage 5: Virtual organizations

At the beginning of the 1990s, American scholars used the network concept to develop the idea of the virtual organization, which emphasizes the disappearing boundaries between organizations (Ashkenas et al., 1995) and the role of information technology. A recent cover story of *Business Week International* described the virtual corporation as follows: "The virtual corporation is a temporary network of independent companies – suppliers, customers, even erstwhile rivals – linked by information technology to share skills, costs, and access to one another's markets . . ." (February, 1993). "This new, evolving, corporate model will be fluid and flexible – a group of collaborators that quickly unite to exploit a specific opportunity. Once the opportunity is met, the venture will, more often than not, disband" (Byrne et al., 1993). Davidow and Malone (1992) paint a similar picture by arguing that "the virtual corporation will appear less a discrete enterprise and more an ever-varying cluster of common activities in the midst of a vast fabric of relationships."

Peters (1994) discusses the same idea when he talks about the corporation as Rolodex. Truett and Barrett (1991) focus on the role of electronic networks, and suggest that Corporate Virtual Workspaces (CVWs) will be highly productive replacements for current work environments, which all but eliminate the need for the physical corporation: "having no need for physical facilities other than the system hosting the CVW, the cyberspace corporation will exist entirely in cyberspace." Loebbecke and Jelassi (1997) describe a case study of how a company transformed itself into a virtual organization. Chesbrough and Teece (1996) point to the vulnerability of a virtual organization, and emphasize the need to match organizational design to the nature of the innovation and the existence of the required capabilities.

Stage 6: Co-opetition

Many firms have noted that the ideal of cooperation may clash with the more traditional philosophy of competition. For instance, writers on the subject of strategic alliances have discussed how companies may safely cooperate with their competitors (e.g. Lewis, 1990). In today's confusing business environment, where other organizations may simultaneously be competitor, customer and partner, companies face the challenge of successfully combining cooperation and competition (Jarillo, 1988; Jorde and Teece, 1989; Bleeke and Ernst, 1993; Faulkner, 1995; Tezuka, 1997). Carlin et al. (1994) call these kind of complex relationships 'multifaceted' and maintain that they are becoming more and more common, especially in high-technology and global industries: "As economies become increasingly global, industry boundaries blur, and technology keeps redefining markets, the situation of having a competitor as a supplier, customer or partner will continue to increase." The combination of cooperation and competition is referred to as *co-opetition* by Lipnack and Stamps (1993). Firms that engage in co-opetition cooperate together to achieve shared goals, but at the same time remain independent from each other. Thus, collaborating competitors may continue to compete in other areas. Brandenburger and Nalebuff (1996) define co-opetition as a revolutionary mindset that combines competition and cooperation and provide numerous examples of firms such as Intel, Nintendo, Ford, Microsoft, American Express and NutraSweet that have used the strategy to dominate their industries.

Internal cooperation revisited

As discussed above, the internal cooperation between the various business functions within a firm may serve as a foundation for different types of external cooperation. Parallel to the evolutionary development described above, the study of internal cooperation has developed in 3 broad directions:

1. *Internal cooperation between two business functions*
   Initially, internal cooperation has been perceived as a coordination problem concerning 2 different business functions. Most notable is the extensive investigation of the

Marketing-R&D interface in developing new products (Gupta et al., 1985, 1986; Souder, 1988; Gupta and Wilemon, 1990; Hise et al., 1990; Song and Parry, 1992; Griffin and Hauser, 1994; Moenaert et al., 1994). In addition, other kinds of internal cooperation have been studied as well, such as the coordination between Marketing and Production and Marketing and Finance (Wilson, 1994).

2. *Internal cooperation between more than two business functions*
Some authors have recognized that a number of organizational processes require the integration of more than 2 business functions. For instance, the successful development of new products depends on the effective coordination of Marketing, R&D and Production (see e.g. Song et al., 1997). As a director of the R&D operations of a food company put it in the mission statement of his business unit, "I have a DREAM: Development by Research, Engineering And Marketing" (cited in Moenaert et al., 1994). Other authors have argued for more involvement of Purchasing in product development (Burt and Soukup, 1985; Williams and Smith, 1990).

3. *Integration of internal and external cooperation*
The 2 evolutionary paths of development come together in the investigation of the relationship between internal and external cooperation. Several companies are effectively blending internal and external cooperation by including key suppliers and major customers in their product development teams. In their report on an extensive study of practical measures to improve product development processes, Howard and Guile (1992) state that "Membership in [cross-functional] teams may be quite flexible and may even include suppliers, customers, consultants and members of technical advisory groups as needed". Such observations have led Biemans (1995) to argue for a more extensive investigation of the relationship between internal and external cooperation. This area has only recently begun to receive the attention it deserves (see e.g. the discussion by Lipnack and Stamps (1997) of virtual teams).

THE FATE OF SMALL FIRMS

The developments presented above are particularly relevant to small firms. For instance, small firms (especially high-tech small firms) tend to depend to a large extent on the input of other organizations. These may be key customers, but more often they are suppliers of key components or part of the required technology to develop new products. In addition, small firms frequently cooperate with universities and commercial research institutions to obtain specific knowledge. One of the problems that very small firms experience in this context is that they are often relatively unfamiliar with the relevant players in the market and what they have to offer. For this reason, the Dutch government, for example, stimulated the development of a complex support structure of information agencies, science parks and innovation centres. The private sector has witnessed the establishment of various knowledge brokers who specialize in specific industries and attempt to bring together potential partners. The various Dutch innovation centres have developed and successfully applied a systematic technique to stimulate and develop partnerships between small firms. Also, leading firms such as the Dutch copier manufacturer Océ van der Grinten encourage cooperation between local, often small firm suppliers, at various levels. Depending on the duration of the cooperation, the resulting collaborative structures display the characteristics of clusters or virtual organizations.

With respect to internal cooperation, small firms frequently experience specific problems related to their size. For example, some critical business functions may be less developed. Marketing may be non-existent or part of Sales, resulting in a short-term focus on immediate sales rather than the long-term development of a sustainable market presence. Time pressures and the ever-present critical issue of survival also contribute to such a short-term orientation. Needless to say, the management teams of small firms are generally quite unaware of the issue of integrating external and internal cooperation, but establish such structures instinctively.

MANAGEMENT WITHIN NETWORKS

Concerning cooperation and networks, there is a large difference between the way the subject is dealt with by academics and practitioners. Academics tend to focus their research on cooperative structures that are typically investigated from a static perspective, characterizing the nature and strength of bonds linking partners, and the measuring of such concepts in terms of distance along various dimensions. Practitioners, on the other hand, try to translate the conceptual idea of networks into concrete activities and changed behaviour, which frequently culminates in equating networking with the active exchange of business cards. For example, Kanter and Eccles (1992) have argued, "whereas academics typically talk about 'networks' or 'network organizations', it is much more common for managers to talk about 'networking'. In contrast to academics, who are interested in *understanding* the noun and adjective, are the managers, who are interested in *using* the verb."

Therefore, having presented the evolution of management thought on organizational cooperation, we will now discuss some of the major issues involved in the management of relationships and networks. These issues can be grouped into 6 broad areas:

1. The strategic decision to cooperate
2. Determining the type of cooperation
3. The selection of a partner
4. The mobilization of key individuals
5. Drawing up an agreement
6. The management of cooperation

Since an exhaustive discussion of all issues involved would require the space of a book, we will limit ourselves to a general discussion, followed by an overview of the most relevant issues for managers.

The strategic decision to cooperate

For many industries and firms, cooperation with other organizations has become critical to survival and continued competitive advantage. As Tom Peters (1994) put it: "Networking is life itself for many leading-edge firms. For MCI, as for Apple, the ability to create – then manage and then disintegrate – networks of assorted sizes and flavors is arguably core competence No. 1." Lipnack and Stamps (1993) put it even stronger when they state that, "Today, no company can go it alone all the time. It's too complicated and too expensive. Right now, it means missed opportunities. Tomorrow, it means going out of business." Although the most frequently cited examples of cooperation involve high-tech industries (particularly the fast moving computer and semiconductor industries), and professional services (with Hollywood film making being the quintessential example), current business practice demonstrates that cooperation is fast becoming a way of life in all kinds of activities, varying from aerospace and petrochemicals to such traditional businesses as publishing, wood-working, furniture and railway freight transportation (Lipnack and Stamps, 1993).

Considering all this, the main issue is not whether to cooperate or not, but rather a set of strategic questions concerning the function and role of cooperation. The major questions management should ask itself in this context are:

- How is cooperation used in our industry (to what extent, with what kind of partners, for which objectives, and for which activities)?
- To what extent are promising partners available in our industry (i.e. not locked into other cooperations and willing to cooperate)?
- What do we want to accomplish through cooperation with other organizations?
- What types of partner do we consider for cooperation (e.g. cooperation with a major customer versus cooperation with suppliers or competitors)?
- With how many different partners do we want to cooperate (specified for every type of partner)?

Determining the type of cooperation

When management is convinced of the merits of cooperation with other organizations and has determined the kind of partner(s) it is looking for (e.g. customers, suppliers, competitors, universities), the next step involves a number of decisions concerning the type of cooperation. Firms may use a large number of different types of cooperation, varying from a joint marketing arrangement to an equity investment or joint venture. Depending on the objectives of cooperation, a firm may select the most appropriate type of cooperation (Huyzer et al., 1990). For instance, when the objective is to acquire new technologies, a firm may opt for (a) sponsoring fundamental research at a number of universities, (b) corporate venturing, (c) a joint venture or (d) a pre-competitive R&D arrangement with a competitor. If, on the other hand the cooperation aims at establishing a presence in a foreign market, the company could choose from (a) a joint venture, (b) a joint marketing or joint selling arrangement with a competitor or supplier of complementary products, (c) a franchise agreement with a local partner. Some types of cooperation are closely linked with specific objectives, while other forms (such as the ubiquitous joint venture) are of a more general nature. Some relevant questions for management are:

- What is the specific objective of our cooperation?
- In which activities do we want to cooperate with partners?
- What specific contributions do we expect from potential partners and what do we have to offer in return?
- What would be the most appropriate kind of partner (e.g. a customer or competitor and should it be a local, domestic or international company)?
- What intensity of interaction do we strive for in our cooperation?

The selection of a partner

A major set of managerial decisions is associated with the selection of an appropriate partner. Numerous authors have joined the debate about the major question of whether one should select a partner on the basis of similarity or complementarity. Hagedoorn (1990) summed up the answer by stating that a firm should aim at similarity balanced by complementarity. Numerous real-life examples illustrate that compatible organizational cultures increase the likelihood of success. However, of more importance is the complementarity of the cooperation partner since the creative combination of complementary activities, knowledge and skills helps realize the desired synergy. Good examples of this kind of synergy are provided by the collaboration between small, innovative biotech companies and large drug manufacturers. These examples also demonstrate that 2 partners need not be of equal size, but should be of equal value to each other (Hull and Slowinski, 1990; Botkin and Matthews, 1992). Several authors have suggested models to evaluate potential cooperation partners (e.g. Souder and

Nassar, 1990a, 1990b). Experienced firms draw up a detailed profile of the ideal candidate before they start looking for a potential partner. A sample of relevant questions for managers might include:

- What is the profile of the ideal cooperation partner?
- How many potential partners should we take into consideration?
- How can we effectively evaluate potential partners, and from where do we get the required information?
- What is the track record concerning the cooperation, experience and expertise of potential partners?
- What is the effect of cooperation with one particular partner on (potential) cooperation with another partner?

The mobilization of key individuals

While the existing literature addresses, in great detail, the selection of the most appropriate partner, the critical importance of identifying and motivating key individuals is very much underestimated. Key individuals should be mobilized at both the partnering firm, and within the partner organization. Nevertheless, there is a difference here in that, while key individuals at the partner organization need to be identified and supported, internal champions tend to be self-appointed.

Many firms have experienced the importance of working through the right individual(s) within a potential partner organization during the initial contact stage. Such a person provides a window on the organization and may supply pertinent information on various aspects of the partner organization. In addition, he or she serves as an internal ambassador by promoting the benefits of the proposed cooperation, building the collective spirit and countering initial resistance, while later on he or she may prove critical in keeping the project going, despite initial setbacks (for this reason such individuals are often referred to as *project champions* (Lewis, 1990)). To be able to function as such an "ambassador", the selected individual(s) need to be provided with "ammunition" (e.g. both compelling qualitative arguments and quantitative data where possible) to counter resistance and demonstrate the benefits of the proposed cooperation. Berry (1980) refers to this process as "managing evidence". Similar roles can be played by key individuals within the partnering organization. In both organizations, the various roles may be played by different individuals, depending on the stage of the project and the base for their authority. For instance, Gemünden (1985) calls them *promotors* and distinguishes between a promotor by use of power and a promotor by use of know-how. Relevant questions for management include:

- What do we want to accomplish by selecting and motivating key individuals?
- Where (i.e. at which level, and in which part of the organization) can we find individuals who would be most effective for our purposes?
- What are the most effective channels of communication for reaching these key individuals and what specific information do they need?
- What is the role of these key individuals over time (e.g. during the initial negotiations, or during the subsequent cooperation project)?
- How can we transfer the experiences of key individuals to the rest of the organization?

The drawing up of an agreement

After having selected the best partner from among the many alternatives available, and achieving verbal agreement to start a cooperative project, detailed agreements concerning a large number of issues need to be arrived at. In addition to clarifying the basis for the collaboration (e.g. clear-cut goals, the contributions of each partner, the division of tasks and responsibilities, the allocation of costs and benefits, confidentiality, the length of the project life, arrangements for termination of the project), the written agreement may also refer to a number of control mechanisms to contribute to a successful collaboration (such as interorganizational decision making, reward systems, motivation of personnel, resolution of conflicts and (in)formal communication processes).

Nevertheless, the role of contracts in strategic alliances is somewhat ambiguous. While they may serve to clarify issues and provide partners with a legal document containing everything agreed upon during negotiations, experienced managers do not consider the contract to be an instrument to extract compliant behaviour from an unwilling partner. Written agreements are especially relevant for specific strategic alliances, such as when no prior alliance relationship exists, and the alliance is seen as a starting point for larger collaborative efforts in the future, or when capital investments are made independently by each of the separate partners within their own companies (Lynch, 1993). The questions that managements need to address include the following:

- For what purposes do we need a written agreement?
- What is the relationship between the written contract and informal agreements?
- Which topics should be addressed in the written agreement and what level of detail should be used?
- How do we prevent the written agreement from promoting contractual paralysis (i.e. a situation where the parties continuously refer to the contract and, as a result, slow down progress)?
- How will the written agreement be adapted to the dynamics of the on-going collaboration?

The management of cooperation

From the very moment the project is begun, effective management of communications (both formal and informal), and recognition of the critical role played by an effective operational interface, become essential ingredients for success. Periodic reviews (at both the strategic and operational level) can be used to measure progress and keep the cooperation project and the relationship on track. However, especially with long-term collaborative efforts, one should always allow for an unexpected "turn of events". In addition, numerous operational differences, such as differences in decision making authority, decision making styles, time orientation and reporting procedures, will only occur during cooperation. Management faces both the challenge and responsibility to employ the creative management of relationships, and thus turn problems into opportunities and apprehension into success. In so doing, a clear focus on the partnership, rather than the well being of the individual firm, leads the way to success.

The cooperation project will be especially successful if the partners develop mechanisms (structures, processes and skills) to bridge interorganizational and interpersonal differences. Such mechanisms should include a flexible interface, with personal contacts at various levels within the organization, while some firms choose to establish a *relationship manager* for the express purpose of coordinating all contacts between both partners. Nevertheless, the appointment of a relationship manager does not absolve employees from their individual responsibility for relationships and for communications (Drucker, 1988). As has been stated before, successful cooperation with other organizations also requires effective internal communications. The questions for management include:

- What are the characteristics of an effective operational interface?

- Which mechanisms can be used to stimulate the transfer of knowledge between partners, and within one's own organization?
- How do we promote the establishment of personal relationships and prevent them from becoming dysfunctional?
- How do we maintain the flexibility to adapt the cooperation to changed circumstances?
- Can the cooperation project be used as a springboard for more extensive collaborative efforts?

Management of cooperation by small firms

All of the issues presented above are relevant for small firms that attempt to cooperate successfully with external partners. Nevertheless, these problems are often exacerbated by time pressures and the absence of the required management knowledge, which prevent them from implementing current insights and best practices. For instance, the selection of partners is more often than not dictated by chance and limited to a few local candidates. Personal relationships and the level of perceived trust play important roles in the development of cooperation between small firms, and frequently eclipse the careful consideration of strategic alternatives.

Lack of experience with such endeavours frequently causes small firms, in general, and high-technology small firms in particular, to make a number of mistakes that are typical of a neophyte at the game of cooperation. Examples include:

- paying insufficient attention to the objectives of the cooperative endeavour;
- neglecting to formulate the specific contributions of both partners;
- 'selecting' the wrong partner (e.g. based on chance contacts or geographical proximity rather than careful consideration);
- neglecting to stimulate and grow the cooperative venture;
- omitting to draw up clear-cut agreements about contributions, the division of benefits, duration of the cooperation and conditions of termination;
- not succeeding in establishing an effective operational interface between both partners (frequently resulting in the loose management approach of solving problems as they occur instead of anticipating them and proactively managing the project);
- failing to fully learn from the collective experiences with the cooperation project so that the knowledge gained can be applied in the future to new situations;
- ignoring opportunities to extend the cooperative project to new ones.

AN AGENDA FOR RESEARCH

Despite the considerable attention that has been paid to cooperation and networking during the last 15 years, a lot of work remains to be performed to further our understanding, and to translate it into practical managerial guidelines. Therefore, without discussing them extensively, we end this paper by suggesting a number of promising avenues for future research.

- *The functioning of firms within a network*
  Since most investigations of networks are rather abstract, much can be learnt from studies that explicitly aim to generate knowledge about the way firms may improve their functioning within a network structure. One major area of research should be focused on how firms may effectively orchestrate multiple types of cooperation.
- *The link between internal and external cooperation*
  As external cooperation with other organizations is closely related to the

interfunctional cooperation within the firm, we need more information about the relationship between these 2 kinds of cooperation. For instance, it is still largely unclear under which circumstances an effective interfunctional cooperation is a prerequisite for effective external cooperation, or to what extent both types of cooperation rely on the same kind of skills and capabilities.
- *The context of a cooperation*
  Most studies of cooperation attempt to generate managerial guidelines that are universally applicable. However, every industry has its own peculiar idiosyncrasies which may influence the occurrence of problems and the efficacy of solutions. Studies in this area need to isolate the relevant industry-specific factors, and determine their influence on cooperation in practice.
- *Human resource management*
  The establishment of cooperation with other organizations requires new skills and capabilities of the people involved. This has far reaching consequences for a firm's human resources management function. Issues to be addressed include the required skills, performance measures, hiring practices, training requirements and reward systems.
- *Combining competition and cooperation*
  As an increasing number of firms realize that they need to combine competition with cooperation, there is a growing need for empirical studies that address the integration of these 2 management perspectives. Much remains unknown about the most effective ways in which to combine competition and cooperation, the influence of this on the selection of partners, and its consequences for the firm's organizational structure (e.g. should a cooperation project with a major competitor be separated from the rest of the organization?).
- *Evolution of cooperation patterns and practices*
  Both empirical research and managerial practice have shown that cooperation is not static and does not last forever. For instance, the large majority of joint ventures are terminated after 6 or 7 years. Thus, a promising area of research might address the way different types of collaborative projects evolve over time, the determinants of success and the means that managers have to shape these evolutionary developments.
- *Specific issues for small and medium-sized firms*
  Although all kinds of firms face the challenge of cooperation with other organizations, we have emphasized that the nature of the cooperation, the specific problems encountered and the appropriateness of various solutions may be strongly determined by the size of the organization. In particular, future research needs to take a closer look at the specific issues that small and medium-sized firms have to deal with.

The accumulation of knowledge is analogous to an expanding body in that it only increases in areas where new insights need to be discovered. This explains why, despite the overwhelming attention paid to cooperation and networks, considerable work remains to be achieved in order to further our understanding, and assist managers in coping with their present dynamic environment.

REFERENCES

Abeele, P. vanden and Christiaens, I. (1987), De Klant als Generator van Innovatie in "High-Tech" Markten – Een Conceptuele en Empirische Studie (The Customer as Generator of Innovation in "High-Tech" Markets – A Conceptual and Empirical Study), *Economisch en Sociaal Tijdschrift*, 1, pp. 27–56.

Anderson, J.C., Håkansson, H. and Johanson, J. (1994), Dyadic Business Relationships Within a Business Network Context, *Journal of Marketing*, 58 (October), pp. 1–15.

Ashkenas, R., Alrich, D., Jick, T. and Kerr, S. (1995), *The Boundaryless Organization: Breaking the Chains of Organizational Structure*, Jossey-Bass, San Francisco.
Asmus, D. and Griffin, J. (1993), Harnessing the Power of Your Suppliers, *The McKinsey Quarterly*, 3, pp. 63–78.
Badaracco, J.L., Jr. (1991), *The Knowledge Link: How Firms Compete Through Strategic Alliances*, Harvard Business School Press, Boston, MA.
Berry, L.L. (1980), Services Marketing is Different, *Business Magazine*, May–June.
Biemans, W.G. (1991), User and Third-Party Involvement in Developing Medical Equipment Innovations, *Technovation*, 11, 3, pp. 163–82.
Biemans, W.G. (1992), *Managing Innovation within Networks*, Routledge, London.
Biemans, W.G. (1995), Internal and External Networks in Product Development – A Case for Integration, in: Bruce, M. and Biemans, W.G. (eds.), *Product Development: Meeting the Challenge of the Design-Marketing Interface*, John Wiley, Chichester, pp. 137–59.
Biemans, W.G. (1996), Organizational Networks: Toward a Cross-Fertilization between Practice and Theory, *Journal of Business Research*, 35, 1, pp. 29–39.
Biemans, W.G. and Brand, M.J. (1995), Reverse Marketing; A Synergy of Purchasing and Relationship Marketing, *International Journal of Purchasing and Materials Management*, 31, 3 (Summer), pp. 29–37.
Biemans, W.G. and Setz, H.J. (1995), Managing New Product Announcements in the Dutch Telecommunications Industry, in: Bruce, M. and Biemans, W.G. (eds.), *Product Development: Meeting the Challenge of the Design-Marketing Interface*, John Wiley, Chichester, pp. 207–29.
Bleeke, J. and Ernst, D. (1993) (eds.), *Collaborating to Compete – Using Strategic Alliances and Acquisitions in the Global Marketplace*, John Wiley, New York.
Booz-Allen & Hamilton (1968), *Management of New Products*, Booz-Allen & Hamilton Inc., New York.
Booz-Allen & Hamilton (1982), *New Products Management for the 1980s*, Booz-Allen & Hamilton Inc., New York.
Botkin, J.W. and Matthews, J.B. (1992), *Winning Combinations – The Coming Wave of Entrepreneurial Partnerships Between Large and Small Companies*, John Wiley, New York.
Brandenburger, A.M. and Nalebuff, B.J. (1996), *Co-opetition*, Doubleday, New York.
Bronder, C. and Pritzl, R. (1992), Developing Strategic Alliances: A Conceptual Framework for Successful Co-operation, *European Management Journal*, 10, 4 (December), pp. 412–21.
Burt, D.N. and Doyle, M.F. (1993), *The American Keiretsu: A Strategic Weapon for Global Competitiveness*, Irwin Professional Publishing, Chicago.
Burt, D.N. and Soukup, W.R. (1985), Purchasing's Role in New Product Development, *Harvard Business Review*, September-October, pp. 90–7.
Burt, R.S. (1982), *Toward a Structural Theory of Action, Network Models of Social Structure, Perception and Action*, The Free Press, New York.
Burt, R.S. and Minor, M.J. (1984), *Applied Network Analysis – A Methodological Introduction*, Sage Publications, Beverly Hills, CA.
Byrne, J.A., Brandt, R. and Port, O. (1993), The Virtual Corporation – The Company of the Future Will Be the Ultimate in Adaptability, *Business Week International*, February 8, pp. 36–40.
Carlin, B.A., Dowling, M.J., Roering, W.D., Wyman, J., Kalinoglou, J. and Clyburn, G. (1994), Sleeping with the Enemy: Doing Business with a Competitor, *Business Horizons*, September-October, pp. 9–15.
Chesbrough, H.W. and Teece, D.J. (1996), Organizing for Innovation: When is Virtual Virtuous?, *Harvard Business Review*, January-February, pp. 65–73.
Ciccantelli, S. and Magidson, J. (1993), Consumer-Idealized Design: Involving Consumers in the Product Development Process, *Journal of Product Innovation Management*, 10, pp. 341–7.
Coleman, J.S., Katz, E. and Menzel, H. (1966), *Medical Innovation: A Diffusion Study*, Bobbs-Merrill, Indianapolis.
Cutts, R.L. (1992), Capitalism in Japan: Cartels and Keiretsu, *Harvard Business Review*, July-August, pp. 48–55.
Davidow, W.H. and Malone, M.S. (1992), *The Virtual Corporation: Structuring and Revitalizing the Corporation for the 21st Century*, HarperCollins, New York.
Devlin, G. and Bleackley, M. (1988), Strategic Alliances – Guidelines for Success, *Long Range Planning*, 21, 5, pp. 18–23.

Drucker, P.F. (1988), The Coming of the New Organization, *Harvard Business Review*, January-February, pp. 45–53.
Dyer, J.H. (1996), Specialized Supplier Networks as a Source of Competitive Advantage: Evidence From the Auto Industry, *Strategic Management Journal*, 17, pp. 271–91.
Dyer, J.H. and Ouchi, W.G. (1993), Japanese-Style Partnerships: Giving Companies a Competitive Edge, *Sloan Management Review*, Fall, pp. 51–63.
Faulkner, D. (1995), *International Strategic Alliances – Co-operating to Compete*, McGraw-Hill, Maidenhead.
Ford, D. (1997), *Understanding Business Markets: Interaction, Relationships and Networks*, The Dryden Press, London.
Foxall, G.R. and Tierney, J.D. (1984), From CAP1 to CAP2: User-Initiated Innovation from the User's Point of View, *Management Decision*, 22, 5, pp. 3–15.
Gattorna, J.L. and Walters, D.W. (1996), *Managing the Supply Chain: A Strategic Perspective*, Macmillan, London.
Gemünden, H.G. (1985), "Promotors" – Key Persons for the Development and Marketing of Innovative Industrial Products, in: *Industrial Marketing – A German-American Perspective*, Backhaus, K. and Wilson, D.T. (eds.), Springer, Berlin, pp. 134–66.
Griffin, A. and Hauser, J.R. (1994), *Integrating Mechanisms for Marketing and R&D*, ISBM Report 14–1994, Pennsylvania State University, University Park, PA.
Gupta, A.K., Raj, S.P. and Wilemon, D. (1985), The R&D-Marketing Interface in High-Technology Firms, *Journal of Product Innovation Management*, 2, pp. 12–24.
Gupta, A.K., Raj, S.P. and Wilemon, D. (1986), A Model for Studying R&D-Marketing Interface in the Product Innovation Process, *Journal of Marketing*, 50 (April), pp. 7–17.
Gupta, A.K. and Wilemon, D. (1990), Improving R&D/Marketing Relations: R&D's Perspective, *R&D Management*, 20, 4, pp. 277–90.
Hagedoorn, J. (1990), Organizational Modes of Inter-Firm Co-operation and Technology Transfer, *Technovation*, 10, 1, pp. 17–29.
Håkansson, H. (ed.) (1982), *International Marketing and Purchasing of Industrial Goods: An Interaction Approach*, John Wiley, Chichester.
Håkansson, H. (ed.) (1987), *Industrial Technological Development: A Network Approach*, Croom Helm, London.
Håkansson, H. and Eriksson, A-K. (1993), Getting Innovations Out of Supplier Networks, *Journal of Business-to-Business Marketing*, 1, 3, pp. 3–34.
Hallén, L. and Johanson, J. (eds.) (1989), *Advances in International Marketing, Vol. 3 – Networks of Relationships in International Industrial Markets*, JAI Press, Greenwich, CT.
Harvard Business Review (1990), Can a Keiretsu Work in America?, September-October, pp. 180–97.
Herstatt, C. and von Hippel, E. (1992), Developing New Product Concepts via the Lead-User Method: A Case Study in a Low-Tech Field, *Journal of Product Innovation Management*, 9, September, pp. 213–21.
Hippel, E. von (1978), Successful Industrial Products from Customer Ideas, *Journal of Marketing*, 42 (January), pp. 39–49.
Hippel, E. von (1986), Lead Users: A Source of Novel Product Concepts, *Management Science*, 32, 7 (July), pp. 791–805.
Hise, R.T., O'Neal, L., Parasuraman, A. and McNeal, J.U. (1990), Marketing/R&D Interaction in New Product Development: Implications for New Product Success Rates, *Journal of Product Innovation Management*, 7, pp. 142–55.
Howard, W.G., Jr. and Guile, B.R. (eds.) (1992), *Profiting from Innovation – The Report of the Three-Year Study from the National Academy of Engineering*, The Free Press, New York.
Hull, F. and Slowinski, E. (1990), Partnering with Technology Entrepreneurs, *Research-Technology Management*, 33, 6 (November-December), pp. 16–20.
Huyzer, S.E., Luimes, W., Spitholt, M.G., Slagter, W.J., van Wijk, A.H., van der Leest, D.J. and Croese, D. (1990), *Strategische Samenwerking*, Coopers & Lybrand Dijker Van Dien, Amsterdam, Samsom BedrijfsInformatie, Alphen aan den Rijn.
Huyzer, S.E., Moëd, J., Spitholt, M.G.M., Luimes, W., Kroodsma, H. and Douma, M.U. (1991), *Strategische Samenwerking; De Praktijk in Nederland*, Coopers & Lybrand Dijker Van Dien, Amsterdam, Universiteit Twente, Faculteit der Technische Bedrijfskunde, Enschede, Samsom BedrijfsInformatie, Alphen aan den Rijn.

James, B.G. (1985), Alliance: The New Strategic Focus, *Long Range Planning*, 18, 3, pp. 76–81.
Jarillo, J.C. (1988), On Strategic Networks, *Strategic Management Journal*, 9, pp. 31–41.
Jorde, T.M. and Teece, D.J. (1989), Competition and Cooperation: Striking the Right Balance, *California Management Review*, Spring, pp. 25–37.
Kanter, R.M. and Eccles, R.G. (1992), Making Network Research Relevant to Practice, in: *Networks and Organizations: Structure, Form, and Action*, N. Nohria and R.G. Eccles (eds.), Harvard Business School Press, Boston, MA, pp. 521–7.
Lazarsfeld, P.F., Berelson, B. and Gaudet, H. (1948), *The People's Choice*, 2nd Edition, Columbia University Press, New York.
Leenders, M.R. and Blenkhorn, D.L. (1988), *Reverse Marketing; The New Buyer-Supplier Relationship*, The Free Press, New York.
Lewis, J.D. (1990), *Partnerships for Profit – Structuring and Managing Strategic Alliances*, The Free Press, New York.
Lipnack, J. and Stamps, J. (1993), *The TeamNet Factor – Bringing the Power of Boundary Crossing into the Heart of Your Business*, Oliver Wight, Essex Junction, VT.
Lipnack, J. and Stamps, J. (1997), *Virtual Teams: Reaching Across Space, Time, and Organizations with Technology*, John Wiley & Sons, New York.
Loebbecke, C. and Jelassi, T. (1997), Concepts and Technologies for Virtual Organizing: The Gerling Journey, *European Management Journal*, 15, 2 (April), pp. 138–46.
Lorange, P. and Roos, J. (1991), Why Some Strategic Alliances Succeed and Others Fail, *The Journal of Business Strategy*, January/February, pp. 25–30.
Lorange, P. and Roos, J. (1992), *Strategic Alliances – Formation, Implementation and Evolution*, Blackwell Publishers, Cambridge, MA.
Lynch, R.P. (1993), *Business Alliances Guide – The Hidden Competitive Weapon*, John Wiley & Sons, New York.
Magnet, M. (1994), The New Golden Rule of Business, *Fortune*, February 21, pp. 28–32.
Millson, M.R., Raj, S.P. and Wilemon, D. (1996), Strategic Partnering for Developing New Products, *Research-Technology Management*, May-June, pp. 41–9.
Miyashita, K. and Russell, D.W. (1994), *Keiretsu: Inside the Hidden Japanese Conglomerates*, McGraw-Hill, New York.
Moenaert, R.K., Souder, W.E., De Meyer, A. and Deschoolmeester, D. (1994), R&D-Marketing Integration Mechanisms, Communication Flows, and Innovation Success, *Journal of Product Innovation Management*, 11, pp. 31–45.
Moore, J.F. (1996), *The Death of Competition: Leadership and Strategy in the Age of Business Ecosystems*, HarperCollins, New York.
Niederkofler, M. (1991), The Evolution of Strategic Alliances: Opportunities for Managerial Influence, *Journal of Business Venturing*, 6, pp. 237–57.
Olsen, R.F. and Ellram, L.M. (1997), A Portfolio Approach to Supplier Relationships, *Industrial Marketing Management*, 26, pp. 101–13.
Parkinson, S.T. (1982), The Role of the User in Successful New Product Development, *R&D Management*, 12, 3, pp. 123–31.
Peters, T. (1994), *The Tom Peters Seminar – Crazy Times Call for Crazy Organizations*, Vintage Books, New York.
Porter, M.E. (1990), *The Competitive Advantage of Nations*, The Free Press, New York.
Richter, F-J. and Wakuta, Y. (1993), Permeable Networks: A Future Option for the European and Japanese Car Industries, *European Management Journal*, 11, 2, pp. 262–7.
Rogers, E.M. (1995), *Diffusion of Innovations*, 4th Edition, The Free Press, New York.
Sabbagh, K. (1996), *Twenty-First-Century Jet; The making and marketing of the Boeing 777*, Scriber, New York.
Saunders, M. (1994), *Strategic Purchasing and Supply Chain Management*, Pitman Publishing, London.
Shaw, B. (1986), *The Role of the Interaction between the Manufacturer and the User in the Technological Innovation Process*, Ph.D. Thesis, Science Policy Research Unit, University of Sussex.
Sheth, J.N. and Sharma, A. (1997), Supplier Relationships: Emerging Issues and Challenges, *Industrial Marketing Management*, 26, pp. 91–100.
Slowinski, G., Farris, G.F. and Jones, D. (1993), Strategic Partnering: Process Instead of Event, *Research-Technology Management*, May-June, pp. 22–5.

Song, X.M., Montoya-Weiss, M.M. and Schmidt, J.B. (1997), Antecedents and Consequences of Cross-Functional Cooperation: A Comparison of R&D, Manufacturing, and Marketing Perspectives, *Journal of Product Innovation Management*, 14, pp. 35–47.

Song, X.M. and Parry, M.E. (1992), The R&D-Marketing Interface in Japanese High-Technology Firms, *Journal of Product Innovation Management*, 9, 2, pp. 91–112.

Sonnenberg, F.K. (1992), Partnering: Entering the Age of Cooperation, *Journal of Business Strategy*, 13, 3 (May/June 1992), pp. 49–52.

Souder, W.E. (1988), Managing Relations Between R&D and Marketing in New Product Development Projects, *Journal of Product Innovation Management*, 5, pp. 6–19.

Souder, W.E. and Nassar, S. (1990a), Choosing an R&D Consortium, *Research-Technology Management*, 33, 2 (March-April), pp. 35–41.

Souder, W.E. and Nassar, S. (1990b), Managing R&D Consortia for Success, *Research-Technology Management*, 33, 5 (September-October), pp. 44–50.

Spekman, R.E., Isabella, L.A., MacAvoy, T.C. and Forbes III, T. (1996), Creating Strategic Alliances Which Endure, *Long Range Planning*, 29, 3, pp. 346–57.

Tezuka, H. (1997), Success as the Source of Failure? Competition and Cooperation in the Japanese Economy, *Sloan Management Review*, 38, 2 (Winter), pp. 83–93.

Truett, S. and Barrett, T. (1991), in: *Cyberspace: First Steps*, M. Bendikt (ed.), MIT Press, Boston.

Voss, C.A. (1985), The Role of Users in the Development of Applications Software, *Journal of Product Innovation Management*, 2, pp. 113–21.

Williams, A.J. and Smith, W.C. (1990), Involving Purchasing in Product Development, *Industrial Marketing Management*, 19, pp. 315–19.

Williamson, O.E. (1975), *Markets and Hierarchies*, The Free Press, New York.

Wilson, I. (ed.) (1994), *Marketing Interfaces: Exploring the Marketing and Business Relationship*, Pitman Publishing, London.

Woodside, A.G. and Wilson, E.J. (1994), Tracing Emergent Networks in Adoptions of New Manufacturing Technologies, 1994 Research Conference on Relationship Marketing, Emory Business School, Emory University, June 11–13, Atlanta.

CHAPTER 3

Science in the Market Place: The Role of the Scientific Entrepreneur

KEITH DICKSON, ANNE-MARIE COLES AND HELEN LAWTON SMITH

INTRODUCTION

At a superficial level it can appear that the roles of the career scientist and small business entrepreneur are diametrically opposed. The highly educated, reflective, cautious scientist has little in common with the risk taking, adventurous industrial pioneer. These stereotypes, however, mask a far more complex phenomenon, whereby active and creative scientists from a variety of institutional backgrounds can identify and attempt to exploit a market niche which they have identified in the course of their research.

The existing literature on the scientist as entrepreneur is as brief as it is confusing. Commentators have tended to concentrate on an analysis of the behaviour of the academic scientist who forges commercial links with industrial partners in order to exploit the results of research. In particular the focus has been on aspects of biomedical science and genetic engineering, as the most obvious areas where academic research has immediate commercial potential (Dubinskas, 1988).

Commercialising academic research is, however, only one side of the story. The crucial role of the entrepreneur acting as a champion of new technology has long been recognised in the innovation literature. The results of project SAPPHO (Robertson et al., 1972) identified a number of entrepreneurial roles within the firm that contributed to success in technological innovation. Scientists working in the research and development division of large corporations act as entrepreneurs inside the company by steering through new and radical technologies that otherwise would not be brought to fruition. Between the extremes of the academic scientist and the industrial technologist lie many possibilities for scientific entrepreneurial activity.

THE ACADEMIC SCIENTIST AS ENTREPRENEUR

In the literature 2 major questions dominate the analysis of the academic entrepreneur. The first is whether the freewheeling academic can come to terms with behavioural constraints demanded by the commercial restrictions on knowledge transfer, and the second concerns the apparent resistance of the academic to acquiring the appropriate business and management skills to successfully develop a small business.

Etzkowitz (1983; 1989) analyses the increasing participation of university-based scientists in commercial ventures. He concludes that these scientists are reacting mainly to changes in the way science is funded which has, in turn, produced changes in the way that scientists work. He found that the commercial sponsorship of research was accepted, and did not cause any particular problems in its execution. He concludes that a transformation of the traditional scientific norms has taken place to accommodate these changes in the source of funding. Far from feeling that this is a deviation from acceptable scientific behaviour, the scientists involved have changed their position to incorporate the idea of capitalising on research as part of their desire to extend the basis of knowledge. Changes in behaviour noted by Etzkowitz include activities such as withholding publication while patents are sought. The academics studied 'become more aware of the potential economic value of their research, and are open to translating research into market products' (Etzkowitz, 1989, p. 27).

These observations relate to the interpretation given by Mulkay (1976) that the norms and values accorded to the scientific method amount less to a rigid prescription about procedure than to a scientific ideology, which leaves individual scientists free to accommodate research within their professional capacity.

Louis et al. (1989) explore the preconditions for entrepreneurial life scientists working in research universities in the USA. They note the growing trend towards commercial activity among university scientists, but conclude that commerce is not commonly entered into by life scientists, and that universities vary greatly in their administrative support for entrepreneurial effort. The authors identify 5 separate types of entrepreneurial activity. These include: large scale projects funded by commercial interests; consultancy work; soliciting funds from industry; patenting; and forming a new company to exploit the results of research. They conclude that life scientists in research universities are moderately entrepreneurial, although they are not often involved in large scale externally funded projects. Entrepreneurial activity is affected by the attitude within the particular institution and entrepreneurialism tends to concentrate in particular universities.

The issues facing academic scientists in the USA are also supported, to a certain extent, by the work of Van Dierdonck and Debackere (1988) through their investigation of academic entrepreneurship at universities in Belgium, despite the different cultural and organisational settings. They studied both the phenomena of academics offering commercial services from the institution, and also of becoming involved in spin-off companies to exploit their own research. The reasons for these activities are identified as: pressure on universities to carry out useful research, the reduction in funding for academic research, the growing inter-relatedness of science and technology in some disciplines, and the desire of academic scientists to become involved in 'real life' projects.

Despite the fact that there is an increasing tendency for academics to become more entrepreneurial, the authors identify a range of barriers which impede the process. These lie in 3 areas. First there are cultural impediments, in that academics and industrialists have different values and expectations, which are not always easy to communicate. Second, institutional inhibitions exist, in that the 2 organisations are not set up to cooperate easily. Third, from an operational viewpoint there is a lack of material support and managerial advice for such ventures. They conclude that academic researchers in Belgium are not sufficiently aware of the entrepreneurial opportunities of their position, and that they lack entrepreneurial behaviour and skills.

There are undoubtedly similar examples of entrepreneurial activity in UK universities. Consultancy work and projects funded by industry are increasing, and the growing linkage between industry and universities has been documented (Lawton Smith, 1991; Charles and Howells, 1992). Moreover, there are the same pressures on academic research funding in the UK as elsewhere, and the responses, in terms of exploring opportunities for exploitation of research, are also similar.

This overview of the academic scientist as entrepreneur illustrates a number of trends. First, it is clear that academic scientists are facing pressures on funding and have been forced to search for sources of commercial funding. Second, while engaged in academic research, scientists are aware of the possible commercial potential of their work. These activities can be accommodated along with their own continuing research interests since the scientists can alter and rationalise their behaviour as a legitimate part of scientific work, and in some cases, a positive impact can be created from becoming involved in real industrial problems. There does not appear necessarily to be any conflict for the scientists involved in undertaking work which has a distinctly non-academic rationale.

ACADEMIC SCIENTISTS AND SMALL BUSINESS

Although a number of entrepreneurial activities are available to academics, the instances of leaving an academic background and setting up a small business is the most clear demonstration of abandoning purely academic goals and taking on the rigours of business. The analysis of the strengths and weaknesses of such undertakings by scientists has received some attention. Samsom and Gurdon (1993) report on a detailed study of 22 cases of scientists who started new businesses in the USA, looking at the background of the scientists and their motivations. The institutional backgrounds of the scientists were a number of academic settings, universities, governmental research facilities and an independent research institutions. They all subsequently established a small business as a result of a research discovery which was perceived to have commercial potential.

In an extended analysis of the fate of these new ventures, when scientists collaborate with management specialists, Samsom (1991) identifies conflicting attitudes between the scientific role and managerial constraints as a clash of cultures. The scientists involved in a technology development tend to take a long term view, while retaining dual loyalty to scientific research and to an entrepreneurial venture, while the management specialists involved took a short term, more commercially focused view. He concludes that scientists commonly fail to appreciate the need for both technical and professional management skills, in the commercial environment. There is a danger that these small firms will become 'technology obsessed' through technological considerations becoming dominant. Market demand and the need for business training are often ignored. Scientists frequently fail to perceive management as a serious discipline, so that,

> comparing the MBA and the scientist cultures shows some real extremes. Neither of these extremes is prepared to really listen, and face the other point of view when they have started a company together. (Samsom, 1991, p. 95)

This research found the scientists to be older than entrepreneurs in general, a fact explained by the longer time spent training to obtain a doctorate. The motivations of the scientists ranged from the desire to have freedom to apply science to real problems in order to advance science, and to become involved in a business. The scientists had a personal acceptance of the challenge involved in the venture, and many had invested their own money, while still *rejecting* the business culture, and preferring to retain links with university research.

The decision for a scientist to set up a small business is one with more far reaching implications. In these cases the scientist is exposed at first hand to the rigours and demands of the commercial sphere, while appearing to retain allegiance to the aims of scientific enquiry. The major difference between this activity and those of the academic scientist is that the exposure is not just to a scientific or technical project that has been defined by industry, but survival demands acceptance of the commercial realities of the market. It is clear that their academic background and scientific achievements do not prepare scientists for the rigours and demands of commercial survival. These businesses are undoubtedly threatened with failure, unless the 2 cultures can be brought closer together.

THE TECHNICAL ENTREPRENEUR

As studies of personnel employed in industrial research and development have illustrated there is not necessarily a 'natural' aversion to business by scientists. Both Kornhauser (1962) and Cotgrove and Box (1970) investigated the integration of scientists and engineers into industry and found they adjusted with little evidence of a clash of interests between their scientific expectations and the commercially oriented projects undertaken by industry. This does not mean that scientists employed in industry do not participate in the traditional activities of knowledge dissemination and exchange. They may also behave as part of a professional community, giving conference papers and publishing articles. Debackere and Rappa (1994) articulate the concept of the technological community, in which both academic and industrial scientists participate. The major point highlighted here is that industrial scientists who hold a doctorate resemble scientists working in academia in their professional conduct.

> As a consequence, simple dichotomies as 'academic' versus 'industrial' may be oversimplified since they do not allow us to fully capture the many subtleties that differentiate technologists in various institutional contexts (Debackere and Rappa, 1994)

A major difference appears to be the possession of a scientific doctorate, since industrial scientists, having only a first degree, do not demonstrate such a preoccupation with professional scientific norms. Thus, the longer scientists have been exposed to an academic culture, the more likely they are to share professional academic concerns. At one level, it is possible to detect this tension between scientific and business cultures, and this has become part of the role of scientists in industry. The scientist may function as part of a business organisation, but is often not perceived as a natural manager. In the extreme, this results in the 'dual ladder', where industrial scientists have a separate career path to that of managers (Allen and Katz, 1986; 1992). The difficulties faced by scientists crossing from R&D management to commercial management are also present in the changing role of research institutions (Turpin and Deville, 1995).

Despite these limitations, R&D scientists are recognised as playing a crucial entrepreneurial role in technical innovation and in championing a new technology through the institutional barriers of large corporations, in order to bring a new development to market. Madique (1980) relates the role played by the entrepreneur in radical innovation to the stage of development of the firm. In a small new firm the technical entrepreneur defines the business, and is often the owner, but in larger corporations they develop a network of differing entrepreneurial roles.

Entrepreneurs are, however, generally associated with a small firm, and the technical entrepreneur is most visible in high technology sectors. In their analysis of the small firm recipients of the UK government SMART awards, Jones-Evans and Steward (1995) identify 4 distinct groups of technical entrepreneur. The 'research' entrepreneur tends to have an academic background; the 'producer' entrepreneur is from an industrial R&D environment, the 'user' entrepreneur has not been directly involved in the development of the technology, while the opportunist has no previous technological experience. Their analysis of these entrepreneurial types illustrates that it is the research entrepreneur who is most likely to have introduced a novel technology, or a new combination of existing technologies. The difference between research entrepreneurs and the more incremental changes made by the others is related to the range of technical competencies found within the small firm.

Further illustration of the technical competence found within such high-tech firms is given by Young and Francis (1991), whose survey of high and low technology firms reported that the founders of high-tech firms were more likely to be college educated, and more likely to obtain scientific information from technical and professional journals and from contacts in universities. For both types of small firm, business opportunities are conditioned by previous work experience and the particular industrial sector involved. The form a new business takes is dependent on the previous experience of the entrepreneur.

It is in the study of small, high technology firms that the concepts of the scientific and the technical entrepreneur meet and overlap. These firms are formed by academic or industrial R&D practitioners who have identified an opportunity to commercially exploit the results of their original research. Such firms become part of a research intensive network, and the ability and success of such firms to spread the costs of innovation by partnering other firms through collaborative R&D projects has been noted elsewhere (Dickson et al., 1997). While the authors emphasise the shift in influence of the small firms with large industrial partners in long standing collaborative projects, Hull and Slowinski (1990) emphasise that becoming a partner to a large firm is, for some entrepreneurs, an attractive means of gaining support for innovation. They also highlight how small firms bring more than technical expertise to the partnership but also contribute 'a significant proportion of the financing, manufacturing, marketing and management' (p. 19).

A review of the literature uncovers a broad sweep of entrepreneurial activity undertaken by scientists. The problems faced by academic scientists in starting a new business are, in part, echoed by the perceived dual role of scientists and engineers in industry. Professional scientific inclinations are seen to be the origin of novel technologies, but are not compatible with the world of hard, commercial decision making. It appears, however, that an academic background is most likely to generate small, innovative high-tech firms. Both the scientific and the technical entrepreneur in this case share characteristics and limitations. The delineation between the two appears to rest entirely on the particular discipline of study. A distinction is difficult to maintain, because most high-tech firms supply either a particular technology or a new technique to a niche market. The inter-related nature of science and technology in a commercial environment, which has already been identified as one of the factors in stimulating scientific entrepreneurship, blurs the distinction between the two. The ability to develop a novel technology at this level is dependent on sound understanding and application of scientific advances.

CASES OF SCIENTIFIC ENTREPRENEURIALISM

The 6 case studies of UK entrepreneurs outlined here all illustrate aspects of entrepreneurialism that have been identified in the literature. In particular, there are 4 areas which can be explored in greater depth including: the impact on scientific behaviour; the methods of searching for new scientific and business opportunities; the process of innovation management; and the success of commercial decision making. These case studies are divided into 3 separate types: academic scientists engaged in entrepreneurial activity while remaining within a university setting; small business owners who have experienced difficulties adapting to business culture, and entrepreneurs who have successfully integrated the demands of a commercial business with the process of technological advance.

The Academic as an Entrepreneur

The first 2 case studies are of university researchers who run a business in their spare time. The first case concerns a firm located within an academic department which carries out research into advanced materials. The objective of the initiative is to generate a source of income to further fund the ongoing academic research. Indeed, some of the academic work itself has the objective of generating patentable results and prototypes that could be commercially exploited through the business, which is independent of the university. This is a two-way process because work is also solicited from industry which tends to be geared towards short term results such as consultancy work and 'trouble shooting'.

One of the advantages of establishing this type of firm is that it opens up opportunities to apply for government and European funds that are not open to academics. The company has, however, received grants from the Department of Trade and Industry (DTI) to exploit patents. The firm has now been in existence for 12 years and one of the key issues in its success is the relevance of the research to industry. This is a rapidly developing area with many opportunities for innovation applicable among a wide range of industries. The business venture has encouraged the academic researchers to become more sensitive to the commercial potential of their work.

This venture appears to be typical of the type of entrepreneurial activity that has been reported in other universities. It has the support of the university, and has been successful in bringing money into the department. Little risk taking is involved, but there is an understanding of what the department has to offer. There appears to be little conflict of culture or behaviour, the 2 sides of the work are seen clearly to enhance and, to a certain extent, to influence each other, without a conflict of demands.

The second example is also of a small business run by 2 academics in their spare time and it is seen by them mainly as a 'hobby' or an adjunct to their academic positions, although they have industrial experience. The firm is a distributor of high-tech instrumentation for the nuclear industry, manufactured in France. Over 10 years, much effort has been put into developing a diverse customer base in the UK and customising the equipment. This is a venture which rests on the expertise of the owners, both in talking to customers, and interpreting their technical requirements.

Although this venture demands a high level of commitment, both in time taken to find customers and in constant communication with the French firm, it can be kept completely separate from academic work. The firm remains a small part time venture fitted in around other commitments, and is focused directly on a market niche. The only time the owners are conscious of the 2 different roles conflicting is when they meet representatives from competitors at conferences.

This venture differs from the first, in that, rather than integrating the 2 roles, they are separated. There does not appear to be a conflict of values or behaviour between the two and the ability to deliver in the business sense depends on a high level of technical competence. In these cases the continuing academic position means that the commercial risks of running a small business are removed, and that the constraints of continually searching for business opportunities and new markets in order to stay solvent are absent. This is matched, in each case, by acceptance of the different values and orientations of the business world, and a certain satisfaction in applying and utilising a high level of technical knowledge to real situations, while retaining a predominately academic perspective.

A Clash of Cultures?

The next 2 case studies concern former academics who left their respective research institutions to become small business owners, but experienced difficulties in moving from one environment to another. The first case concerns a firm formed by 4 academics who initially set up a part time venture, although it was 5 years before the first full time employee was appointed. The business revolves around high performance lasers, searching out and adapting the technology to suit particular market niches. It depends on continual investment in research and development, a good team spirit, and a high level of commitment from the directors. The staff are research oriented, attend conferences and keep up to date with advances in the field.

From the beginning there were problems for this firm in adjusting to a commercial culture in which knowledge has a commercial value. The directors had to learn that there are differences between university customers, where it is possible to be free and open with information, and commercial customers, where their know-how had to be protected. The other problem involved in adapting to function as a commercial venture has been in learning to understand the market. In many of their joint development ventures with other companies they had left the role of developing the market to the partner. This had not worked, and major investments had not given the return expected. The firm has now a well-developed policy of retaining control over both the direction of technological development and the final marketing of the technology.

The other problem of being a small high-tech company surviving independently in a world market is that it has been subject to the vagaries of global economics. The firm lost vital markets during the last decade and has been obliged to contract to about half its size. Not only have they learnt of the risks of commercial survival, the secrecy and suspicion needed in dealing with other companies, but they must keep abreast of a rapidly changing field to remain a supplier to the leading edge of research. Survival in this field has demanded an understanding and adaptation to a business culture far removed from the research of an academic laboratory.

The next case is of a small electronics firm formed in 1978 by a doctoral graduate in physics who had been working in a government research laboratory. Undercapitalisation was a major problem for many years, and this forced the firm into taking on 'one-off' project work resulting in a rather hand-to-mouth existence. New opportunities are now found through close contacts with industry and universities, attending conferences and exhibitions and a strong policy of academic publications to ensure a high academic profile. The owner's strategy for business development has been to develop scientific contacts in key organisations, to act in a creative, problem solving capacity at the research end of R&D, to consciously interact with research scientists as a research scientist, and to leave the question of funding and project support to the scientists within their organisation.

This has led to 2 difficulties. The firm was vulnerable to large firms trying to control developments to which they had contributed. Also large one-off project collaborations with a large corporation have tied up the entire research capacity of such a small firm. The owner's strategy of interacting as a research scientist without the luxury of established institutional backing has led to a number of cash-flow problems and underinvestment over a number of years. Collaborating with large partners also puts them at the mercy of major changes in R&D strategy made by the large firm, where whole areas of research effort can be abandoned.

These difficulties finally began to recede after a non-executive financial director was appointed to help run the business on more commercial lines. This resulted in changes to the method of selecting new projects, and more forward planning. At the same time new scientific contacts forged by the owner gave rise to 2 fruitful collaborations which resulted in the development of a product line, taking the company into manufacturing on a modest scale.

Success of the firm rests on the ability of the owner to offer technical solutions particularly to large firms working at the leading edge of electronics development. Contacts are made through reputation in the scientific field, and through networking. The customers are research scientists in commercial and academic institutions world wide. New customers come from recommendations gained through the scientific network, and through citations in the academic literature. New projects are built on the ability to offer beyond 'state of the art', prototype solutions to explore new research problems.

The split between the demands on the business for financial survival through careful project selection and the pioneering scientific spirit of the owner has allowed it to evolve and develop into a more stable firm, whose type of business demands a necessary preoccupation with constant innovation. This is perhaps an extreme case where the tensions between business and scientific activity are apparent, which arise from the fact that the firm trades on its scientific professionalism.

Science as a Business

The final 2 cases demonstrate how tensions between the desire to create new knowledge and the need to be business oriented can be resolved. The first example is of a small company which designs, manufactures and distributes instruments for research in molecular biology and genetic engineering. The founder of the company has both a degree in cell biology and a business degree. The business is directed by scientific intelligence in identifying emerging opportunities for products as scientific research areas develop. Business openings are often found by keeping in touch with research trends, attending exhibitions and conferences, and developing and maintaining networks of good links with academics who are often the source of ideas for new products. This is a commercially driven approach to innovation, which does not mean that the company will not back a product that fails, but that business decisions make it possible to change course, and to aim a product at a particular market.

In this case, the driving force for the business is again technological innovation. Products that are developed have a short lifetime due to the speed of scientific development. This, and the fact that the actual products depend mainly on engineering design, mean that the electronic technology is not high level, and is easily copied, making innovation rather than patenting the means of protection. The company was founded on the idea of finding niche markets in molecular biology, a rapidly developing field, identifying emerging areas and quickly producing relevant equipment.

This company appears to trade in a very similar fashion as the previous case. Both are dependent on awareness of the leading edge of research, and are providing mainly enabling instrumentation. The difference lies in how they regard science as a business. This firm tends to be more cautious, market driven and commercially hardheaded. The firm is clearly driven by commercial rather than scientific priorities, and the technologies are not regarded as an end in themselves. This is the only example where the owner has both formal scientific and business qualifications.

The final case is of a small firm which develops bespoke business software. The owner has a doctorate and is expert, particularly in computer graphics and information retrieval. This company was started with venture capital in 1984, which demanded that there were managers as well as strategic software developers involved at the start. The company has remained at the leading edge of software development, but it works on 'one-off' developments in collaboration with clients.

The company remains in touch with developments elsewhere through conferences, journals and exhibitions. The aim is to target niche markets which have a particular need for specialist software. The clients provide the user requirements and the firm develops the requisite software. Each business is different but the high level technical software skills have developed to the extent that each new case can be assessed quite accurately. Negotiations in any new situation are routine, and legal contracts are standard. This firm has been business driven, working at the extreme edge of software development for business use. It has, however, remained a small firm which is expressly interested in leading edge developments in software, and is now entering the multimedia field.

There has been a balance between the core business and more challenging novel developments. New business opportunities are assessed both in terms of the technical challenge offered in software development and in the advantage they will offer to the client. The firm has not been expecting to grow in size but to survive and invest in emerging areas which present real intellectual challenge. In new uncertain areas the company tends to enter into joint ventures with another company in order to spread the risks involved. The company designs technical solutions to business problems, which ensures a steady core business, and provides the opportunity to participate in challenging projects. This remains a small, specialist, technically advanced company, quite unusual in the UK. It has a good reputation, and new customers are often found through the recommendations of satisfied clients.

This company has a clear balance between commercial and technical interests. Maintaining a balance between strictly business negotiations and the leading edge developments supports its continuing survival. Although the owner is careful about the business opportunities with which he is becoming involved, and in particular he prefers to fund ventures which have a high level of technical uncertainty, it is clear that the firm is driven by his intellectual curiosity and desire to remain involved in the emerging developments at the leading edge of software engineering.

CONCLUSIONS

These UK case studies lend some support to the academic analysis that scientists have difficulty in reconciling research demands with business imperatives, but they also demonstrate, to varying degrees, ways in which this apparent tension can be surmounted. In doing so, they also indicate that the issue of the scientist as an entrepreneur is more complex than a simple trade-off between scientific norms and business behaviour. Nevertheless, one can perceive a transition from the purely academic scientist, casually involved in commercial activity, to the business entrepreneur exploiting science. A simple typology can be derived from the case studies with the following categories: academic entrepreneur, entrepreneurial scientist, and scientific entrepreneur.

In the first phase, academic scientists engage in entrepreneurial endeavours, but view it only as an adjunct to their academic work. They do not accept, nor need to accept, the full implications and restrictions involved in running a small business, and can maintain their identity as academic scientists. Those involved in this phase can be termed academic entrepreneurs.

In the second phase, as exemplified in the above section suggesting a clash of cultures, the scientist is operating full-time in a business venture whilst still essentially dedicated to scientific interests. A possible clash of cultures might be perceived, although the individuals concerned might adopt many different priorities in pursuing their business without compromising their scientific values. For example, the owner of the small electronics firm described above exhibited similar antipathy to business norms as the North American entrepreneurs described by Samsom, but over time discovered that the way to link scientific behaviour and business constraints was both to work closely with a financial adviser and to make clear decisions over project selection. The scientist/owner can maintain his scientific interests, while the firm trades on his scientific knowledge and understanding and maintains its long term viability by harnessing this to business planning. Those active in this second phase can best be described as entrepreneurial scientists.

The third phase signals a clear integration of scientific and business interests. In the previous section above, the owner of the biological instruments firm has both scientific and business qualifications, which has enabled him to make crucial decisions about the direction of the company while utilising a high level scientific intelligence to identify new business opportunities. Such scientific entrepreneurs have clearly begun to see science as a business.

Small, innovative, technology-intensive firms, often founded by individuals described above, are an important element of future economic activity and growth, and should be encouraged. It is the scientific entrepreneur who has developed the scientific knowledge and the business acumen to contribute to this growth. To achieve such a balance requires some recognition that scientists endeavouring to become entrepreneurs need some commercial education, if only to lessen the possible tension between scientific and business pursuits. While it is not necessarily a foregone conclusion that scientific and business cultures will clash, there is a danger that, without an adequate understanding of commercial and business constraints, the scientist will fail as an entrepreneur.

ACKNOWLEDGEMENT

The authors acknowledge the financial support of the UK Economic and Social Research Council which funded research on 'Critical Factors in Inter-firm Research Collaboration' (Grant No. L323253014) from which this paper was written.

REFERENCES

Allen, T. and Katz, R. (1986) The Dual Ladder: Motivational Solution or Managerial Delusion?, *R&D Management*, 16, 2, pp. 185–97.
Allen, T. and Katz, R. (1992) Age, Education and the Technical Ladder, *IEEE Transactions on Engineering Management*, 39, 3, pp. 245–73.
Charles, D. and Howells, J. (1992) *Technology Transfer in Europe*, Belhaven Press, London.
Cotgrove, S. and Box, S. (1970) *Science, Industry and Society; Studies in the Sociology of Science*, George Allen and Unwin, London.
Debackere, K. and Rappa, M. (1994) Technological Communities and the Diffusion of Knowledge: A Replication and Validation, *R&D Management*, 24, 4, pp. 355–71.
Dickson, K., Coles, A-M. and Lawton Smith, H. (1997) Staying The Course: Strategic Collaboration for Small, High-Tech Firms, *Small Business and Enterprise Development*, 4, 1, March, pp. 13–20.
Van Dierdonck, R. and Debackere, K. (1988) Academic Entrepreneurship at Belgian Universities, *R&D Management*, 18, 4, pp. 341–53.
Dubinskas, F. (1988) Janus Organisations: Scientists and Managers in Genetic Engineering Firms in Dubinskas, F. (ed.) *Making Time, Ethnographies of High Technology Organisations*, Temple University Press, Philadelphia.
Etzkowitz, H. (1983) Entrepreneurial Scientists and Entrepreneurial Universities in American Academic Science, *Minerva*, 21, pp. 198–233.

Etzkowitz, H. (1989) Entrepreneurial Science in the Academy: A Case of the Transformation of Norms, *Social Problems*, 36, 1, pp. 14–27.
Hull, F. and Slowinski, E. (1990) Partnering With Technology Entrepreneurs, *Research Technology Management*, 33, 6, pp. 16–20.
Jones-Evans, D. and Steward, F. (1995) Technology, Entrepreneurship and the Small Firm, *Proceedings of the European Conference on Technology Management*, University of Aston, Birmingham, July 5–7, pp. 272–79.
Kornhauser, W. (1962) *Scientists in Industry*, University of California Press, Berkeley, CA.
Lawton Smith, H. (1991) Industry and Academic Links: The Case of Oxford University, *Environment and Planning C: Government and Policy*, 9, pp. 403–16.
Louis, K., Blumenthal, D., Gluck, M. and Stoto, M. (1989) Entrepreneurs in Academe: An Exploration of Behaviour Among Life Scientists, *Administrative Science Quarterly*, 34, pp. 110–31.
Madique, M. (1980) Entrepreneurs, Champions and Technological Innovation, *Sloan Management Review*, 21, Winter, pp. 59–76.
Mulkay, M. (1976) Norms and Ideology in Science, *Social Science Information*, 15, 4/5, pp. 637–56.
Robertson, A., Achilladelis, B. and Jervis, P. (1972) *Success and Failure in Industrial Innovation: Report on the Project SAPPHO*, Centre for the Study of Industrial Innovation, London.
Samsom K. (1991) *Scientists as Entrepreneurs: Organisational Performance in Science Started New Ventures*, Kluwer Academic Publishers, Boston, MA.
Samsom, K. and Gurdon, M. (1993) University Scientists as Entrepreneurs: a Special Case of Technology Transfer and High-tech Venturing, *Technovation*, 13, 2, pp. 63–71.
Turpin, T. and Deville, A. (1995) Occupational Roles and Expectations of Research Scientists and Research Managers in Scientific Research Institutions, *R&D Management*, 25, 2, pp. 141–57.
Young, R. and Francis, J. (1991) Entrepreneurship and Innovation in Small Manufacturing Firms, *Social Science Quarterly*, 72, 1, pp. 149–62.

PART III Characteristics of HTSF-Entrepreneurs and Their Companies

CHAPTER 4

Management Styles and Excellence: Different Ways to Business-Success in European SMEs

WIM DURING AND MARCO KERKHOF

INTRODUCTION

The research presented here is based upon a joint effort to provide empirical evidence from cases throughout Europe for the proceedings of the 25th European Small Business Seminar (ESBS) of the European Foundation for Management Development (EFMD). The main aim of the European research project was to identify key management attributes which lead to business success in the case of small and medium sized enterprises (SMEs), and to explore the management development challenges facing SMEs (Panayides, 1995). The main dimension of the ESBS analysis is the comparison of empirical results on the areas of management organisation and control, strategic behaviour and orientation, networking, internationalisation and management development between the different countries of Europe. Within the Dutch part of the ESBS research, we have explored newly emerging issues, with a view to providing some insights on different management styles and excellence within successful Dutch companies (During and Kerkhof, 1995).

In this paper we will build upon this theme, and present our findings looking from an European perspective. Only data from manufacturing companies are used as an example of small firms in a technological environment. In line with our thoughts and the issues we discussed in 1995 (During and Kerkhof, 1995) we will look more closely here at the characteristics of innovativeness, and explore further some ideas on the learning styles of the companies involved.

METHODOLOGY

As mentioned before, the research was part of a cooperative project between different European countries, and therefore the methods and instruments for data gathering were kept simple and uniform to the highest possible degree. The main instrument used to gather the necessary information for the research was a structured personal interview questionnaire put to the owner/manager or the general manager. The methodology was based on the European survey of 'successful' firms which were selected to meet pre-defined performance criteria. The main objective was to select high performers belonging to middle-sector companies employing between approximately 10 and 150 persons. For the purpose of this project 2 guiding principles for selecting high performers are an average annual rate of turnover growth of 3% and an average rate of return on equity of 10%. For this paper, only the data from manufacturing companies (n = 227), as an example of small firms in a technological environment, were used.

AN INTRODUCTION OF THE EUROPEAN COMPANIES

In this section some interesting results from the data analysis of general information on survey firms are presented briefly to form a basis for the discussion of issues beginning to emerge from the research. First, some results are presented with respect to the characteristics of the companies involved in the research. The results with respect to the characteristics of the managers of these companies are presented, and finally, some aspects of the strategy and organisational characteristics of the involved companies are displayed. The results are presented in so-called morphological form, which enables us to gain quick insights into the different characteristics of the successful SMEs, and to systematise the findings regarding these characteristics.

Table 4.1 shows that over 70% of the companies involved in the research were family-controlled. Within the research, this lop-sided balance is also seen with respect to the size of the organisation, hereby defined as small businesses employing a maximum of 50 employees, and medium-sized businesses containing between 50 and 250 employees.

Table 4.1: *Characteristics of the companies involved in the research (n = 227)*

• Year of Establishment	in the years 1940–1945 12%	in the years 1946–1980 47%		in the years 1980–1989 40%
• Status	Sole Proprietor 12%	Partnership 17%	Private Ltd 59%	Public Ltd 11%
• Type of Ownership	Family Owned 58%	Family Controlled Shareholders 14%	Multiple Individual Shareholders 20%	Multiple Corporate Shareholders 7%
• Employment (fte, average 1993)	1–20 persons 19%	21–50 persons 66%	51–100 persons 12%	more than 100 3%

Table 4.2 shows that there are several reasons mentioned by the owner/manager for forming his own business. The need for independence is mentioned most frequently, followed by family influence, or a good business idea. In the Dutch research (not included in the Table), the number of jobs the manager had before was also examined. One to 3 jobs scored highest at 60%, followed by more than 3 jobs at 26%.

Table 4.2: *Characteristics of the company managers involved in the research (n = 227)*

• Position	Owner/manager 39%	Share-owner/manager 48%		General manager 13%
• Education	Primary 9%	Secondary 16%	Higher 27%	University 29%
• Reason to run own business	Need for independence 34%	Good business idea 16%	Family 27%	Other reasons 24%

Analysis of Table 4.3 shows that 53% of the companies involved had a formal strategic plan. In the Dutch research, in which 81% of the companies involved had a strategic plan, the data suggest that the plan was communicated to the other managers in 85% of the companies involved, whereas in only 26% of cases was the plan also communicated to the employees.

Table 4.3: *Characteristics of strategy and planning of the companies involved (n = 227)*

• Do you have a strategic plan?	A written strategic plan 53%		No written strategic plan 47%	
• Major strengths (most mentioned)	Quality of products 73%	Quality of staff 7%		Customer care 8%
• Major weaknesses (most mentioned)	Financial recourses 25%	Costs 12%		Quality of staff 13%
• Life cycle of the business	Phase 1: New business 10%	Phase 2: Fast growth business 18%	Phase 3: Consolidation, no growth 24%	Phase 4: After consolidation growth 49%
• Business goal (most mentioned)	Grow fast 12%	Productivity 33%	Minimise costs 19%	Improve quality 15%
• Cooperation within areas	Development 14%	Purchase 18%	Subcontracting 23%	Distribution 21%
• Financial indicators in use	Profit/Loss 23%	ROI 12%	Turnover ratios 20%	Cash-flow 33%
• Number of training days (in 1993)	0–5 days 65%	6–10 days 18%		more than 10 days 18%

INNOVATIVENESS

Introduction

A much debated research issue concerning SMEs and excellence is the link between innovativeness and characteristics of the organisation (see for instance Burns and Stalker, 1961; Rogers, 1983). This is an interesting area of research nowadays, because high tech small firms are being confronted with a number of intertwined changes to their environments. Markets are making more stringent demands, competition is increasing, and in effect, companies simultaneously are required to meet demands for efficiency, quality and flexibility (Abernathy et al., 1981; Bolwijn and Kumpe, 1990). These above described changes in environments have made the successful development and introduction of new products and services a key to success in many organisations.

Therefore, within the dataset of high performing SMEs, we decided to look more closely to the innovative entrepreneurs and their characteristics as examples of high tech small firms.

Innovativeness is defined as 'the degree to which an individual entity is relatively early in adopting and implementing new ideas than other entities of his social system' (adapted from Rogers, 1983). Within this research, organisations which have proactively introduced new products ahead of their competitors are defined to be more innovative than organisations in which new ideas were led by their competitors. In common with other research in this area, this analysis is based on the assumption that an organisation, which is defined to be innovative, must have several distinguishing characteristics. These characteristics can be divided into individual characteristics and organisational characteristics (see for instance Rogers, 1983).

Individual characteristics

In the literature, considerable attention has been paid to the human component of innovativeness (see for instance Schon, 1963). Although his work was not specifically concerned with innovativeness, Newman (1973) has formulated some useful arguments regarding the influence the manager may exert on his work. The relationship between the person and his job is affected by factors such as personality, attitudes and skills. Similar mechanisms are likely to hold for innovative efforts, meaning that the way the innovation process will actually unfold depends on the personal characteristics of the manager.

Within the constraints of this research we decided to look more closely at the way in which the manager can be characterised by his education, his attitude towards the job, and the intensity of systematic training given to managers and employees.

The first proposition is that innovativeness is influenced by the way in which the manager can be characterised by his education and his attitude towards the job, in the way that managers within more innovative organisations are educated in practice, and how they describe themselves as entrepreneurs with creative ideas. However, managers within less innovative organisations, are educated more theoretically and describe themselves as professional managers.

The analysis of the European research data showed a relevant difference between the more and less innovative organisations in terms of the manager's education. More than 25% of the managers in more innovative organisations had obtained post-graduate or professional qualifications, whereas 85% of the managers in less innovative organisations had not received a university degree.

Furthermore, the analysis of the Dutch research data showed another, less significant, difference between the more and less innovative organisations regarding characterisation of the manager's work. Roughly 75% of the managers in more innovative organisations described themselves as entrepreneurs, against about 50% of the managers in less innovative organisations.

In general, education of the high tech manager and orientation towards entrepreneurship may be important elements in the propensity for innovativeness. In innovation processes, organisation members face uncertainty, and strategies for dealing with these situations usually have not been covered by pre-established rules and procedures. Here greater reliance must be placed on informal networks and interpersonal relationships.

It is also expected that innovativeness is influenced by the intensity of systematic training given to managers and employees. Managers within more innovative organisations are willing to invest more time and effort in acquiring knowledge on management in general and on innovation opportunities and constraints more specifically, whereas managers in less innovative organisations are less willing to invest their time and effort.

The analysis of the European research data showed a relevant difference between the more or less innovative organisations regarding the intensity of training. Of the managers in more innovative organisations, roughly 20% accumulate more than 10 training days per year, while almost 25% attended between 6 and 10 days of training. However, almost 75% of the managers in less innovative organisations attended less than 5 days of training per year.

In general, if the organisation's knowledge of, and experience with, innovative efforts are insufficient to meet the requirements of the innovation process, knowledge gaps are bound to occur during the process. One way to cope with this is a strategy aimed at diminishing knowledge gaps by increasing the organisation member's innovative capabilities by means of training and education over a wide range of competencies, including technical and organisational issues, leadership, motivation and communication.

Organisational characteristics

It is generally accepted that innovativeness is influenced by organisational characteristics, and several authors have studied structural and cultural differences to find out what it is that makes some organisations more progressive or innovative than others (see for instance Burns and Stalker, 1961; Rothwell, 1977). Among the many organisational characteristics that have been identified as affecting innovativeness, in the literature 3 variables have received considerable attention, namely: centralisation, formalisation and professionalism (see for instance Zaltman et al., 1973). There is a fair degree of convergence in the literature concerning the direction in which innovativeness is affected by these 3 dimensions. Low degrees of centralisation and formalisation, and a high degree of professionalism, are conducive to the early stages of the innovation process, whereas the opposite conditions facilitate the later stages. Furthermore, it is accepted that the characteristics of the innovation process change in the course of time, requiring different forms of organisation during this process. Therefore, an innovative organisation should also incorporate flexibility, in order to be able to face uncertainty. Flexibility can be obtained by the single organisation internally and/or externally between organisations within a production chain, or network.

Within the constraints of this research we decided to look more closely at the perceived strengths and weaknesses and the growth strategy of the organisation. We also examined the way flexibility is incorporated within the organisation in terms of the way it is structured, and whether the organisation is collaborating with other firms.

It was expected that there would be a difference between more and less innovative organisations with respect to their growth strategies and their perceived strengths and weaknesses.

The analysis of the European research data showed a relevant difference between the more or less innovative organisations regarding growth strategy. Although 60% of respondents in both groups mention market expansion as their main growth strategy, in almost 30% of the more innovative organisations, growth strategy was based on quality improvements, whereas in more than 20% of the less innovative organisations growth strategy was based on increasing the number of products. By relating innovativeness to market expansion, a considerable difference can be noted with respect to entering new markets. Within the 60% of both groups who grow by market expansion, a considerable difference can be noted with respect to the entering of new markets. Market expansion for the more innovative companies is at over 80% as opposed to only 60% for the less innovative companies.

The analysis of the European research data showed no difference between the more or less innovative organisations with regard to their first mentioned strong point. Roughly 70% in both groups mention quality as their first strength. Nevertheless, a considerable difference can be noted between the more or less innovative organisations regarding their second strengths. In almost 30% of the more innovative organisations new ideas or marketing are mentioned as second strength, whereas in the less innovative organisations these strengths were not mentioned. Indeed, in almost 55% of the less innovative organisations customer care was mentioned as a second strength, whereas this was mentioned by only 30% of the more innovative organisations.

In general, detailed awareness of the organisations' products may be an important element in triggering innovativeness. Attention must be focused on generating new ideas to develop new products, marketing of the new product, and the improvement of product quality.

It was expected that innovativeness would be influenced by the way the organisation was structured, and by the way that more innovative organisations have employed a high number of skilled professionals in a project-oriented approach, with decentralisation of responsibility and authority. However, less innovative organisations, which tended not to focus on the above structural changes, were expected to have employed more unskilled employees in their organisations.

The analysis of the European research data showed a considerable difference between the more or less innovative organisations. Over 20% of the more innovative organisations had employed over 20% of skilled professionals, whereas only 10% of the less innovative organisations have employed this number of skilled employees. Also, only about 20% of the more innovative organisations had employed over 20% of unskilled employees, whereas almost 45% of the less innovative organisations had employed this number of unskilled employees.

The analysis of the Dutch research data showed a considerable difference between the more and less innovative organisations. In the past 5 years, 44% of the more innovative organisations had changed towards a project-oriented organisational approach, whereas only 10% of the less innovative organisations had adopted this regime. There is also a slight difference with relation to the changes towards a higher degree of decentralisation of responsibilities and authorities in that 53% of the more innovative organisations had changed, against 40% of the less innovative organisations.

In general, considering that innovation is on a relatively small scale, but is a continuous activity within innovative organisations, the use of project groups to integrate the activities of functional teams responsible for specialist problem solving will (in general) be the most viable arrangement. An advantage of functional grouping is that group members may learn from each other and become more adept in their work (see for instance Mintzberg, 1979a). However, a disadvantage of this approach is that such groups lack mechanisms for coordinating the workflow. Therefore, the task of the project team is to deal with workflow interdependencies, whereas the functional teams are designed to cope with process interdependencies.

Another variable which is expected to influence innovativeness is the attitude of firms towards collaboration with other firms in the area of product design and development. Although operational cooperation (with higher efficiency) usually is the main focus for organisations regarding collaboration with other firms, cooperation can also be aimed at developing new products and services (see for instance Biemans, 1989; Bradburry, 1989; Kerkhof and Boer, 1995). By means of interactive relationships, organisations can shorten the duration of the total product development process, obtain the necessary knowledge, and share the costs and risks involved. Therefore, it might be expected that innovative organisations also are more willing to focus on strategic collaboration with other firms, and on the area of cooperative product design and development. Less innovative organisations, however, might tend to focus on operational cooperation.

Nonetheless, based on the analysis of the European research data, no difference can be found between the more and less innovative organisations on strategic collaboration. Although our analysis did not show any statistically significant relationship, there is a slight tendency for more innovative organisations to favour strategic collaboration. Almost 10% of the more innovative organisations were focused on collaboration with other firms within the area of cooperative product design and development, against about 5% of the less innovative organisations.

Detailed analysis of the European data showed a significant difference in operational cooperation areas. In 50% of the more innovative organisations, operational cooperation was focused on 'subcontracting out' and cooperative use of production, whereas in 80% of the less innovative organisations operational cooperation was focused on cooperative purchase, cooperative distribution and acting as a subcontractor. Further analysis of the subcontracting relationship showed that the more innovative organisations more often operate as placers of subcontract work (almost 80%), whereas less innovative organisations more often act as performers of subcontracts (almost 60%).

In general, the most effective strategy to cope with the complexity of innovation, which is dependent on its newness and radicalness, is to involve a sufficiently wide range of experts within the process. Knowledge gaps are bound to occur if the members of the organisation are not able to meet specific requirements of the innovation process. One way to cope with this is through training (see above), while another way is to collaborate with members of other organisations, on a permanent basis, or for the duration (or part) of the innovation process. Strategic collaboration can be aimed at developing new products and services, but in most organisations it takes quite some time to develop this strategic view and implement it in practice (see for instance Kerkhof and Boer, 1995).

Summarising characteristics of innovativeness

This research has shown a tendency for managerial characteristics to be different between the more and less innovative organisations. Managers of innovative organisations describe themselves as entrepreneurs, and invest more time and effort in systematic training.

The research also showed a difference in organisational characteristics between more and less innovative organisations. No tenable relationship could be noticed between more and less innovative organisations with respect to their willingness to focus on strategic collaboration with other firms within the area of cooperative product design and development. However, a relevant difference between the more and less innovative organisations was noted with respect to areas of operational cooperation. Innovative organisations more often operated as principal placers of subcontract work, whereas less innovative organisations more often act as providing subcontractors. Also a considerable difference existed between the more and less innovative organisations with respect to their internal organisations. Innovative organisations employ more skilled professionals, and less unskilled employees. They have also changed towards project-oriented organisations, with a decentralisation of responsibility and authority in order to be flexible and cope with uncertainty.

In general, the organisation should respond flexibly to evolving circumstances, and apply the most suitable strategy. In this respect, considering that coping with uncertainty may be utilised positively through learning, Schroeder et al. (1986) suggest that understanding the learning process may be a better method of (innovation) management than attempting to remove all setbacks and surprises. A view of organisational learning styles is discussed in the next paragraph.

Organisational learning styles

As previously stated, companies are nowadays operating in a world full of turbulence and rapid change. They have to anticipate and adapt to these changes, partly by radical innovation, and partly by continuous improvement. Successful companies find ways of coping with these changes, demanded from outside and inside their organisations.

The ability to continuously adapt to change is characteristic of what is called 'a learning organisation'. Descriptions of the learning organisation are generally rather normative, as if there is just one best way of organising. This is contrary to the well established contingency approaches of organisation theory. One well documented innovation contingency within the industrial small business is the management style of the business-manager, or entrepreneur (During, 1986). For the purposes of this paper we analysed our data to obtain better insights into the concept of the learning organisation.

In the Dutch database we found some interesting indications of different learning-styles for younger and older managers. The European database contains less detailed information, so we can only execute a partial comparison with the Dutch data. However, the findings are interesting enough to suggest scope for further research into the learning-styles of SMEs.

March and Olsen (1975) and Argyris and Schön (1978) were among the first researchers to introduce the notion of a learning organisation. They viewed learning, in the first place, as another way of looking at the world, of thinking about the world, and acting in the world. The learning organisation, as they see it, can be characterised in the following way:

- *People* are seen as motivated to learn. Under the right conditions they are willing to develop their abilities and improve their performance,
- *The organisational structure* is comprised of small parts, with clear and complete tasks. Decisions are made by the people who actually fulfil the different tasks,
- *The organisational culture* is one which continuously questions the ways in which things are achieved, and if improvements are possible in the light of concrete experience.

In his, by now famous, book 'the fifth discipline' Peter Senge proposes 5 ingredients of a learning organisation (Senge, 1990). These can be summarised as follows (Dumaine, 1994):

- people should put aside their old ways of thinking (mental models),
- they should learn to be open with others (personal mastery),
- they should understand how their company really works (systems thinking),
- they should form a plan upon which everyone can agree (shared vision),
- and they should work together to achieve a common vision (team learning).

In our experience it is doubtful that all successful SMEs conform to the rather normative descriptions presented above. As is well documented in the literature, the characteristics of SMEs are largely determined by the characteristics and management style of their top manager(s) since they have a stake in all major decisions within the organisation (Scherjon, 1994). We propose that different companies have found different ways to fulfil their learning needs, and that what Senge calls a learning organisation represents a specific learning style. Furthermore, we propose that these learning styles are in some way related to the characteristics of top manager.

Because of its roots in many disciplines such as psychology, economics, management science, social science and cybernetics, the concept of organisational learning is rather ambiguous (Wijnhoven, 1995). As a common ground in many of these different perspectives, Wijnhoven finds that organisational learning is mediated by individual learning in an organisational context (containing structures, procedures, norms, culture and information systems). The result is a change in organisationally shared knowledge. Shrivastata (1983) also states that organisational learning is based in individual learning by arguing that: 'since individuals are the agents of learning, their role in the development and institutionalisation of learning systems is critical'. In modelling the organisational learning process several other authors have taken the individual learning cycle of Kolb (1974) as their guide (see for instance Carlsson et al., 1976; During, 1986; Schein, 1993; Kim, 1994). According to Kolb, different individuals develop different learning styles, which is a proposition that is related to the concept of cognitive style (Vollers,

1980; Boekaerts, 1978). If individuals develop distinct learning styles, and organisational learning is mediated by individuals, it is not an exaggeration to suppose that organisations also develop a specific learning style. Indeed Shrivastata (1983) presents a typology of, what he calls, organisational learning systems, based on literature about organisational learning and empirical studies of organisational learning. Shrivastata uses 2 dimensions to construct his typology as follows:

> On the *individual-organisational dimension*, organisational learning systems may vary from being single person dependent systems to highly participative ones, depending on how the process of knowledge sharing is accomplished. The learning systems which are based on inputs from one or a few critical individuals are usually biased by the personal preferences and values of these individual members.

> On the *evolutionary-design dimension*, organisational learning systems are However, learning systems may develop on a purely evolutionary basis. Socio-cultural norms and historical practices lead to a kind of conventional wisdom about how certain problems should be solved, and by whom. At the other extreme, organisational learning systems are designed and implemented to serve specific information and learning needs, identified by managers. Decision support systems, which are often computer based, are special cases of this kind of learning system (p. 15).

Shrivastata identifies 6 organisational learning systems within this framework, based on a study of organisational learning systems in 32 businesses. Three of these systems are typical for large or purely professional organisations, while the other 3 are of interest for the analysis of our SME sample.

The one man institution
One person (i.e. the peak co-ordinator [see Mintzberg, 1979b]) is the key to organisational learning. He is considered to be the source of all critical information. Referring to Argyris and Schön (1978), Shrivastata states that the perceptions and ambitions of this person become organisational perceptions and limitations through a process of tacit acceptance by organisation members.

Participative learning systems
This system refers to the practice of forming teams of all sorts for resolving strategic and management control problems. Key organisation members are in face to face contact, enabling the development of the understanding of each others assumptions and perspectives on critical issues. In its pure form this type of decision making tends to be protracted.

Formal management systems
These systems refer to established procedures for dealing with standard, and some non-standard, activities. Examples comprise strategy development systems, environmental scanning systems and financial and budgetary control systems. These systems incorporate learning from organisational experience, and learning from modern management theory and management techniques.

Learning styles in practice

Based on the theory and prescriptions of organisational learning presented above, we expect that the use of training, the use of external expertise, the delegation of authority and decision making, and the leadership style will show distinct differences for the 3 learning systems presented here. By analysing these variables, a difference between the companies of older managers (over 50 years of age) and younger managers (50 years of age or younger) with respect to the organisational learning style have become apparent.

Training (related to personal mastery, teamwork/team learning, systems thinking)
On the whole older and younger managers spend about equal amounts of time on training. There is a slight difference in participation in short courses, (i.e. 83% for older managers as opposed to 80% for younger managers). In both groups it is a way to develop their personal skills and their insights into new developments in, for example, the marketplace or technology.

The Dutch data show longer training periods for younger managers, which may imply they are more oriented towards modern management theory as opposed to the more issue-oriented training attractive to older managers.

Use of external expertise (related to specific questions and to more generic questions of external developments affecting mental models and personal mastery/craftsmanship)
It is generally accepted that Dutch managers are relative 'heavy users' of external consultants. Our data confirm this picture, since, in the European sample-group 30% of the younger managers and 43% of the older managers do not use external consultants. However, for the Dutch data the figures are 14% and 29% respectively. The findings suggest that Dutch firms use consultants more widely while, within the Dutch sample, younger Dutch managers are most willing to change their mental models.

Of those who work with consultants, 23% of the younger and 28% of the older manager groups do so in the field of finance. (In the Dutch data the comparable figures were 20% for both). For human resources these figures were 14% and 12% respectively, for the European sample (while in the Dutch data the comparative figures were 32% versus 21%). However, in the field of general management, with 18% versus 12% for the European sample, younger managers use outside consultants more frequently then older managers, while the comparative Dutch data produced a 41%–25% respective balance.

Apparently younger managers place more emphasis on the use of external knowledge for developing new strategies and activities than their older counterparts.

Payment
We expected that older managers would rely more on financial incentives to stimulate the 'right' activities of their employees than older managers. However, this expectation was not supported by our results regarding level of payment. In our sample, 70% of the older managers pay more than the level of collective agreements, whereas only 55% of the younger managers did so.

But another explanation for this result might be that older managers may have more long term higher paid employees in their companies.

In addition to the European questionnaire, we put additional questions to the Dutch sample on the delegation of authority and leadership style.

Delegation of authority/internal entrepreneurship (including the use of clear objectives and delegated budgetary authority; a clear picture of the relatedness of tasks to overall performance; accountability and learning from experience)
Of both management age groups, about 65% of senior managers in each category developed their business plans in co-operation with a wider management team. This is a means of using the expertise of all members of the team to develop a 'shared vision'.

Older managers made more use of additional assistance from accountants and family contacts. Possibly, their role in decision making in the management team was more pronounced.

Financial budgets were much in use in all companies. Within the boundaries of the budget, decision authority was delegated to operational levels. Younger managers, however, made more use of budgets, in 77% of cases against 63% of older managers' companies.

Leadership style (joint policy making/team learning)
An important part of a participation type of leadership style is to let employees bring forward information and plans themselves. It involves encouraging initiative, and using the information they gather as the basis for action plans, and if necessary, correcting proposals to conform with company activities as a whole. The use of information about market developments provided by sales people is an indication of the management style in this respect. We found that, of the younger managers, over half used information provided by sales people as their most important source, against less than a third of the older managers.

COMPARING MANAGEMENT STYLES

From the above analysis, we have observed some likenesses and some differences in the management styles of older and younger managers, resulting in different learning styles for their organisations.

What is similar?

Both management groups are result-oriented, they set goals and budgets, they make use of financial reports, and they use financial consultants. These are characteristics of a 'formal management systems' learning style.

Both management groups develop business plans in co-operation with their management team. In this way they secure the involvement of all managers, they make well-informed decisions and they develop a 'shared vision' for the company. This is typical for a 'participative learning systems' learning style.

Both management groups develop themselves by participating in management training activities. This is the case for all learning styles.

What is different?

Younger managers make more use of external consultants for assistance with general management problems, which is indicative of the 'formal management systems' learning style. But consultants are also used for human resource management consultancy, which is an indication of the 'participative learning systems' learning style.

Older managers more often use their own views and insights as the basis for organisational learning, which is indicative of the 'one man institution' learning style.

Younger managers make more use of external trainers for in-company training programmes, which is consistent with both the 'formal management systems' and the 'participative learning systems' learning style approaches.

Again, apparently older managers put more emphasis on existing practices and their own views as guiding principles for the learning behaviour in their companies, in keeping with the 'one man institution' learning style form of management.

The younger manager leadership style is more objective-oriented and delegating (by indicating goals to accomplish), compatible with the 'formal management systems' learning style, as opposed to the more task-oriented (what to do) leadership style of the older managers which is more compatible with the 'one man institution' learning style.

Younger managers make decisions, to a greater degree, on the basis of information provided by employees, than do older managers. This reflects a 'participative learning system' learning style for the younger manager, but a 'one man institution' learning style for the older managers.

Summarising the characteristics of organisational learning styles

The results presented above give some evidence of the existence of 2 different learning styles for the companies of older and younger managers.

The organisational learning style of the older managers' companies is primarily a mixture of what Shrivastata calls the 'one man institution' and the 'formal management systems'.

The organisational learning style of the younger managers' companies is a mixture of 'formal management systems' and 'participative learning systems'.

CONCLUSIONS AND ISSUES FOR FURTHER RESEARCH

In normative management literature, being an active-innovative organisation, and being a learning organisation, are prerequisites to success. On the basis of our research into successful SMEs, this picture needs some adjustment.

With respect to innovation, we found a clear distinction between 2 groups, one being active innovators, the other being passive innovators. We identified some differences in management characteristics and organisational characteristics between these groups. In the active innovating companies the managers see themselves as more entrepreneurial, are practically educated, and invest more time and effort in training. By employing more skilled professionals and less unskilled employees, the organisation is able to transform itself into a more project oriented structure, with decentralised decision taking. To cope with uncertainty, active innovative organisations focus on 'subcontracting out' and the cooperative use of production.

In designing and delivering training and consulting programs to foster success, these differences should be taken into account.

The learning organisation of normative management literature represents one out of several different organisational learning styles. In our sample, we found evidence of 2 different learning styles: one more manager dominated (primarily found with older managers), and the other more participative (primarily found with younger managers).

It is interesting to note that both styles have several characteristics in common. By emphasising these commonalties important requirements for success seem to be met. In other respects, managers may mould their organisation according to their abilities and personal style.

Again these differences should probably have an impact on the way training and consulting programs for success are administered.

Our findings are tentative but of sufficient interest to prompt further research with larger groups of successful SMEs. Further questions might explore whether the relationship between success and differences in innovativeness can be confirmed, as well as the differences in management characteristics. The same interest would hold for the differences in organisational learning styles.

Another question is how these differences should influence the way in which management training and consulting for success in SMEs is conducted.

More research is also needed to find out how innovativeness, management characteristics and organisational learning style are connected.

REFERENCES

Abernathy, W.J., Clark, K.B. and Kantrow, A.M. (1981) The new industrial competition, *Harvard Business Review*, September-October, pp. 68–79.
Argyris, C. and Schon, D.A. (1978) *Organisational Learning: A Theory of Action Perspective*, Addison-Wesley, Reading, MA.
Biemans, W.G. (1989) Developing innovations within networks – with an application to the Dutch medical equipment industry, thesis Technical University of Eindhoven, Eindhoven.
Boekaerts, M. (1978) *Towards a Theory of Learning based on Individual Differences*, Katholieke Hogeschool Tilburg, Tilburg.
Bolwijn, P.T. and Kumpe, T. (1990) Manufacturing in the 1990s – Productivity, flexibility and innovation, *Long Range Planning*, 23, 4, pp. 44–57.
Bradburry, J.A.A. (1989) *Product Innovation: Idea to Exploitation*, John Wiley, Chichester.
Burns, T. and Stalker, G.M. (1961) *The Management of Innovation*, Tavistock, London.
Carlsson, B., Keane, P. and Martin, J.B. (1976) R&D Organisations as Learning Systems, *Sloan Management Review*, 17, pp. 1–15.
Dumaine, B. (1994) Mr. Learning Organisation, *Fortune*, 21, pp. 75–81.
During, W.E. (1986) (in Dutch) *Innovatieproblematiek in kleine industriele bedrijven*, Van Gorcum, Assen/Maastricht.
During, W.E., en Kerkhof, M. (1995) Management styles and excellence, different ways to business success in Dutch SMEs, proceedings of the 25th European Small Business Seminar, Limassol, Cyprus.
Kerkhof, M. and Boer, H. (1995) Networking in transportation and distribution: from operational to strategic collaboration, in Pawar, K.S. (ed.) Proceedings of the Second International Symposium on Logistics, pp. 73–8, University of Nottingham, Nottingham.
Kim, D.H. (1994) (in Dutch) Het verband tussen individueel leren en het leren van organisaties, *Holland Management Review*, 38, pp. 104–16.
Kolb, D. (1974) Learning and Problem Solving: On Management and the Learning Process, in Kolb, D.A., Rubin, I.M. and McIntyre, J.M., *Organisational Psychology: a Book of Readings*, Second Edition, Prentice Hall, Englewood Cliffs, NJ.
Lovelock, Chr. H. (1985) Developing and implementing new services, in George, W.R. and Marshall, C.E. (eds.) *Developing new services*, pp. 44–64, American Marketing Association, Chicago.
March, J.G. and Olsen, J.P. (1975) The uncertainty of the past: organisational learning under ambiguity, *European Journal of Political Research*, 3, pp. 141–71.
Mintzberg, H. (1979a) *The Structuring of Organisations*, Prentice Hall, Englewood Cliffs, NJ.
Mintzberg, H. (1979b) Organisational power and goals: a skeletal theory, in Schendel, D.E. and Hofer, D.W. (eds) *Strategic Management: A New View of Policy and Planning*, Little Brown, Boston.
Newman, D. (1973) *Organisation Design: An Analytic Approach to the Structuring of Organisations*, Arnold, London.
Panayides, G. (1995) *PROMEESE, profiles of management excellence in European small enterprises*, Nicosia, Cyprus.
Rogers, E.M. (1983) *Diffusion of Innovation*, Free Press, New York.
Rothwell, R. (1977) The characteristics of successful innovators and technically successful firms (with some comments on innovation research, *R&D Management*, 7, 3, pp. 191–206.
Schon, D.A. (1963) Champions for radical new inventions, *Harvard Business Review*, March-April, pp. 77–86.
Schon, D.A. (1967) *Technology and Change*, Delta Books, New York.
Schein, E.H. (1993) (in Dutch) Hoe organisaties sneller kunnen leren, *Holland Management Review*, 35.
Scherjon, D.P. (1994) (in Dutch) *Strategisch management in het MKB*, Deventer.
Schroeder, R., van de Ven, A.H., Scuder, G. and Polley, D. (1986) Managing innovation and change processes: findings from the Minnesota Innovation Research Program, *Agribusiness*, 2, 4, pp. 501–23.
Senge, P.M. (1990) *The Fifth Discipline: The Art and Practice of the Learning Organisation*, Century Business, New York.

Shrivastata, P. (1983) A typology of organisational learning systems, *Journal of Management Studies*, 20, 1, pp. 7–28.
Vollers, T. (1980) (in Dutch) De toepasing van een dynamisch innovatiemodel, Masters Thesis, University of Twente, Enschede.
Wijnhoven, A.B.J.M. (1995) *Organisational Learning and Information Systems: the Case of Monitoring Information and Control Systems in Machine Bureaucratic Systems*, University of Twente, Enschede.
Zaltman, G., Duncan, R. and Holbeck, J. (1973) *Innovations and Organisations*, Wiley, New York.

PART IV Entrepreneurial Networking

CHAPTER 5

Preconditions and Patterns of Entrepreneurial Networks in an Innovative Environment

KARL-HEINZ SCHMIDT

PROBLEMS, DEFINITIONS AND DATA

'Network economics' is of growing interest in an innovative environment because changing environmental conditions, and new production systems, need more flexible productive resources and entrepreneurial decision-making (Dosi et al., 1988; Johannson et al., 1994). In order to explore the impact of entrepreneurial networks in an innovative environment, this paper investigates the preconditions and patterns of co-operating customers and suppliers in the face of the differing requirements of technological change, with special consideration of the behaviour of entrepreneurs, managers and employees in high-technology oriented companies and network promoting institutions (Stoneman, 1987). The problems to be analysed concern especially the operation of networks related to subcontracting business, schooling (i.e. courses at vocational training schools) and training (i.e. practical work training 'on the job'), technology transfer and the impact of social and cultural organisations (Teubal and Zuscovitch, 1994). For all of these types of networks the impact of co-operative entrepreneurial activities, and of the co-operative behaviour of employees regarding the diffusion of innovations, is examined.

The definition of the entrepreneurial network is based on the co-operative organisation of companies and on the economic behaviour of entrepreneurs and managers. The co-operative organisation concerns not only the arrangement of the work places and production process in the companies, but also the technological and economic relations of the individual companies within co-operating groups, and the way these groups operate within industrial complexes. The preconditions of entrepreneurial networks include co-operative individual behaviour and the determinant factors of economic activities including productive resources, new technologies and institutional factors.

Depending on alternative sets of preconditions, patterns of entrepreneurial networks differ. Four types of patterns can be distinguished which comprise subcontracting business, co-operative economic interfirm relations, human interaction based on non-economic relations, and the relations of private-public or semi-public institutions.

The flexibility and creativity of entrepreneurs, managers and employees at the company-level are of increasing importance for the successful competitiveness of the enterprises under the conditions of rapid technological change. Therefore the size of the company, measured in terms of employees, invested capital and/or turnover, is relevant to the economic success of entrepreneurial networks. According to a definition applied in the European Union, small and medium enterprises are judged to employ less than 500 persons, while large enterprises employ more than 500 persons (Hofstetter von Trachselwald, 1990; Schmalen and Pechtl, 1992).

The innovative environment of the enterprise in this paper is considered through the analysis of technological changes, technological spillovers, spin-offs and new ventures. The internal structures of survey enterprises are also investigated in terms of company functions such as training, research and development, production, finance, marketing and sales.

The data of this investigation refer to surveys and case studies in the manufacturing sector, and especially in high-technology industries. The following case studies were carried out in different countries, mainly in Japan, Switzerland and Germany, but to some extent also in countries characterised by different technological environments, such as South Africa (Schmidt, 1989; 1994; Pett, 1994). However, survey data mainly refer to EU countries.

PRECONDITIONS OF ENTREPRENEURIAL NETWORKS

First, co-operative individual behaviour is clearly a precondition of entrepreneurial networks through growing entrepreneurial co-operation. Various reasons for this growth can be identified:

- the decentralising effects of new technologies, based on the later stages of micro-electronics production,
- an increasing degree of uncertainty over the economic and technological environment of the enterprises,
- the changing preferences of clients in the market place, and the integration of worldwide markets.

Second, the non-economic relations of individuals or groups of persons are part of the preconditions of entrepreneurial networks.

Third, specific productive resources, including qualified labour, a flexible supply of risk-capital, open access to information on new resources, the results of R&D, new markets and extent of future demand, are part of the preconditions of such networks.

Special emphasis is given in high-technology companies to related network institutions when compared to low-technology companies. According to the results of this author's case studies of manufacturing companies in Japan, Switzerland, Germany and South Africa (Schmidt, 1989; 1990a and b), several types of co-operative entrepreneurs can be distinguished comprising: (A) autonomous (presidential), (B) innovative, (C) imitative and (D) traditional entrepreneurs.

Autonomous entrepreneurs are either aggressive innovators, or they are interested in the stabilisation of their influence on management and production in the company. Innovative entrepreneurs are the 'pioneers' of the economy (Stratos Group, 1990). They are interested in the introduction and diffusion of 'new combinations' in Schumpeter's view, and in the perspectives of New Schumpeterianism (Schumpeter, 1942; Nelson and Winter, 1982; Witt, 1990). Schumpeter distinguished new products, methods of production, resources, purchasing markets and selling markets as 'new combinations', while he terms inventions, innovations and diffusion as phases of the innovation process. This new Schumpeterianism emphasises attitudes and market structure. Imitative entrepreneurs are interested in saving the costs of R&D by rent-seeking. They are risk-averse, and hesitate when considering the introduction of new technologies. Traditional entrepreneurs apply well-known methods of production and try to keep their economic position in a specific (local or regional) market segment. They are mainly found in low-technology companies and in technologically 'old' industries, although these industries may also include the pioneer and autonomous entrepreneurs noted above.

Empirical examples of all the types of entrepreneurs were identified in the considered countries, but entrepreneurial networks turned out to be less widespread in these countries. They were found to some extent (judged by numbers of participating enterprises, employees, production volumes and sales) in industries structured by high rates of subcontracting business, technology transfer, schooling and training and organisations oriented to social and cultural purposes. Specific cases revealing the existence of entrepreneurial networks were found in Japanese manufacturing, mainly in the automobile industry and machine-tool industry. Parts of both industries are characterised by high-technology subcontracting business and export-intensive marketing and sales operations (Schmidt, 1994; 1995). In these industries, entrepreneurial organisations co-operate intensively with local government authorities, prefectural institutions, federal ministries and bureaucracies in order to develop new networks for the training of management and employees of the member companies. The development of entrepreneurial networks is occurring to different extents in European countries, and on a smaller scale, but lately at an increasing rate in South Africa (Schmidt, 1988; 1996). These statements prompt an investigation into the patterns of entrepreneurial networks in the considered countries, especially in innovative industries.

PATTERNS OF ENTREPRENEURIAL NETWORKS

The problems and opportunities of entrepreneurial networks are exemplified when considering the activities in enterprises within the manufacturing industries of Japan, Switzerland and Germany. The fields of activities concern: (I) training, and research and development, (II) production, (III) finance and (IV) marketing and sales. For every field of activity, and for each type of entrepreneurial network, the opportunities of network operation can be pointed out. Case studies of manufacturing companies allow for the description of factors hindering or favouring the establishment of entrepreneurial networks. By use of a matrix for each country, the factors determining the patterns and the development of networks may be exposed.

An example concerning the machine-tool industry in Japan is presented in Table 5.1. It shows that the preconditions and patterns for entrepreneurial networks in that industry have favoured the establishment and expansion of networks organised as subcontracting business, training centres, technology transfer institutions and social and cultural organisations.

As a consequence, relationships aiming at the introduction or growth of subcontracting business function as entrepreneurial networks. In Japan the vertical (hierarchical) structure of subcontracting systems is predominant, whereas in Germany, Switzerland and other European countries flat pyramid or horizontal structures of the subcontracting business are more widespread (Pett, 1994; Schmidt, 1990a and b; 1991; 1994).

In the Japanese machine-tool industry the most important pattern of entrepreneurial network is the subcontracting business, especially the 'pyramid-type' of subcontracting. It consists of a contracting or 'mother' company on top of a pyramid of subcontractor firms, involving different layers of subcontractors, ranked according to the complexity of the products and services which they deliver to the contracting company. The network activities are oriented mainly towards the organisation of the production process, but they also concern the provision of training and R&D. On-the-job-training (OJT), quality-control-circles (QCs), and training courses run by contractors and subcontractors, public and semi-public information centres and counselling services, but organised by Chambers of Industry and supervised by MITI-authorities, generally

Table 5.1: *Patterns of entrepreneurial networks concerning the machine-tool industry of Japan 1994/95*

Patterns of entrepreneurial networks	Fields of network activities			
	Schooling, training, R&D	Production	Finance	Marketing and sales
Subcontracting business	Internal schooling, OJT, QC, internal training of subcontractors, internal R&D	'Pyramid type' subcontracting, OJT, QC, joint R&D of orderers and subcontractors	Joint committees of companies, banks and state officials; internal management groups including 'house banks'	Internal marketing management of subcontracting systems; joint sales offices; joint market research
Co-operative economic interfirm relations	OJT, joint schooling and training, labour transfers, co-operative R&D	OJT, co-ordinated production management, co-ordinated technology management	Co-operative arrangements between companies, banks, insurance and government agencies	Co-operation covering specific market segments; joint marketing activities
Private, public, semi-public institutions	Internal schooling, off-the-job-schooling, MITI, Chambers of Commerce, Trade Unions	Co-ordinated plant location; MITI, Chambers of Commerce, R&D-institutes	Strategic alliances; public information centres; public subsidies for R&D	Private and public marketing organisations, public research institutes, joint information services
Human interaction based on non-economic relations	Discussion-groups, social clubs, cultural organisations	Trade unions, social clubs, cultural organisations	Public relations activities of banks and insurance companies, co-operatives movement	Consumers organisations, social clubs, cultural organisations, environment protection movement

characterise Japanese networking systems. Furthermore they are characterised by the close co-operation of managers, engineers, administrative personnel and workers at the shop-floor level. Although central management is responsible for the planning and financing of investment, networking systems increase the information and motivation of workers, and subsequently their productivity. Diverse information channels, discussion groups, quality-control-circles and labour transfer committees enlarge the motivation of employees and intensify the 'bottom-up' information and proposals developed by employees at the shop-floor level (Ogawa, 1984; 1993).

Compared to the co-operative entrepreneurial networks of high-technology industries in Japan the entrepreneurial networks of European countries are more explicitly oriented to the targets of individual companies. Also, they are less flexible regarding employment and the transfer of labour than the Japanese networking systems. Instead, European networking systems are more strictly oriented to vocational career patterns and to the continuous accumulation of human capital, especially of qualified labour.

The entrepreneurial networks in Switzerland, Germany and other European countries have to endure the rigidities of vocational education and employment practices, but they benefit from synergistic effects of both systems. The high-technology industries especially exploit the synergistic effects of a vocational education system by combining different locations of schooling, training and practical work. But basic problems of this fragmented system of vocational education have increased since the acceleration of technological change and the required reforms of the vocational education system are lagging behind. The curricula have been changed only to a small extent. The continuing growth of the high-technology industries in Central Europe has demonstrated that innovative enterprises need skilled, creative and flexible employees, both at the shop-floor and at business administration levels. Employees must be continuously upgraded in terms of skills. This effect can be achieved by the activities of entrepreneurial networks. Furthermore, it has become obvious that innovative companies need more risk-bearing capital, and open access to information on new technologies and new management-methods.

The impact of entrepreneurial networks on the vocational education and employment systems can be exposed by comparison of Japan, Central Europe and South Africa. In Japan and Europe the entrepreneurial networks contribute to vocational education and to the efficient allocation of highly qualified labour. However, in the 'dual economy' of South Africa, entrepreneurial networks have less opportunities to diffuse innovations or to create new jobs in high-technology industries. Yet, in specific industries including, for example, metallurgic and inorganic materials, chemicals and transportation equipment, advanced technologies are applied also in South Africa, that is in the 'first world' sector. Consequently, the entrepreneurial networks of South Africa reflect a 'dual economic structure' involving high-technology oriented activities in the advanced industries versus low-technology oriented activities in the less developed industries concerned with textiles, furniture, and construction. But additional employment is also created in these industries when they increase production volume. In fact the production volume of low-technology industries has increased in South Africa, mainly through investment in low-technology equipment. The entrepreneurial networks of these industries are focused on the organisation and control of basic vocational training and schooling. Moreover they are oriented to intrafirm and interfirm co-operation concerning information, organisation of procurement, training, financing, production and marketing.

EMPIRICAL DATA OF ENTREPRENEURIAL NETWORKS

The results of case studies of manufacturing industries in Japan, South Africa, Switzerland and Germany performed by this author presume that innovative industries may create new work places, but that this effect can be over-compensated for the loss of jobs elsewhere (Dose and Drexler, 1988). Table 5.2 points out a list of employment creating innovative industries, although their characteristics are not common for every industry in every country. The industry of metallurgic and inorganic materials, for example, is designated as an innovative and employment creating industry in Japan and South Africa, but not in Switzerland and Germany. The electronics industry is recorded as innovative and employment creating in Japan and Central Europe, but not in South Africa. Although these descriptions are based on case studies which consider the management strategies, innovative activities and changes of employees in selected companies, they conclude that entrepreneurial networks have differing effects on employment in the investigated industries in different countries. Especially in the high-technology industries, the innovative environment of the enterprises must be considered. Herewith, the diversity of private, public and semi-public networks must be taken into account. Also, technological spillovers, spin-offs and co-operative networks integrating new ventures and 'adult' enterprises should be analysed. Yet, since the availability of relevant statistics

Table 5.2: *Employment creating innovative industries in Japan (J), South Africa (SA), Switzerland (CH) and Germany (D)*

Industries	Countries			
	J	SA	CH	D
Mechanical engineering	x		x	x
Electronics	x		x	x
Chemicals	x	x	x	x
Metallurgic and inorganic materials	x	x		
Transportation equipment	x	x		x
Automobiles	x	x		x
Energy production	x		x	x
Textiles	x	x	x	
Furniture		x	x	x
Construction		x		
Medicine and pharmaceutical appliances			x	x
Communication equipment	x		x	x
Food production		x	x	
Domestic appliances		x	x	

Source: Own case studies 1989–1996

and survey data is restricted, the analysis of the entrepreneurial networks must be based on different data sources, including own case studies.

In Japan the entrepreneurial networks are part of the institutional framework of industrial organisation and government policies (Lakshmanan, 1994). Private, public and semi-public institutions are involved in co-operations targeted at specific policy programmes or in the supply of informations and counselling to individual enterprises, entrepreneurial organisations, unions or public agencies. Concerning R&D and Technology Transfer (TTr) of high-technology industries, the co-operation of government agencies, universities, science laboratories, research institutes, information and counselling centres, technology parks and science cities or science parks is relevant for the functioning of entrepreneurial networks in this field (Schmidt, 1984; 1990a and b; 1994; 1995).

In Switzerland private organisations, semi-public institutions and, to a smaller extent than in Japan, public institutions take part in the development of entrepreneurial networks. In high-technology industries, the entrepreneurial networks were established mainly through activities of large enterprises, aimed at the diffusion of new technologies and interfirm co-operation.

In Germany the functioning of entrepreneurial networks in high-technology industries is based on the activities of large enterprises, semi-public and public institutions, with a higher impact of public institutions financed by government than in Switzerland. Technology transfer institutions, including high-technology centres and R&D-laboratories financed by the state, contribute to the diffusion of new technologies. They co-operate with institutions financed by entrepreneurial organisations or by semi-public institutions like the Chambers of Industry and Trade (Baranowski and Groß, 1994).

In South Africa private and public institutions participate in the establishment of entrepreneurial networks and public information services and technology transfer. Yet, compared to Japan and Central Europe, the preconditions for the development of high-technology industries are more restricted in South Africa than in Japan, Switzerland and Germany.

Concerning the instruments of the entrepreneurial networks in the high-technology industries of the considered countries the case studies point out different combinations and effects. In Japan, the activities of entrepreneurial networks are, to a large extent, concentrated on science parks and suburban areas in, or near to, agglomerations. In Switzerland and Germany decentralised technology parks, high-technology laboratories, schooling and training institutions and technology transfer centres utilise decentralising instruments, especially information and counselling services, experiments in laboratories, training courses and seminars. In South Africa, the schooling and training courses for low-skilled workers are mainly of an applied nature, but in high-technology industries, the activities of technology parks, laboratories and university related R&D institutes are expanding, especially within and nearby industrial agglomerations.

The empirical data demonstrate the co-operation achieved by institutions in sharing the instruments of R&D, technology transfer, schooling, training and information in high-technology industries and in the industries linked with the high-tech-sector (Acs and Audretsch, 1992; Pohl, 1995). Spin-offs by the establishment of new companies through the spillover effects of the economic and technological relations of 'adult' and 'young' enterprises at different locations are documented by case studies and surveys.

Table 5.3: *Fields of technology applied by enterprises which are located in centres for innovation and establishment of new enterprises in Germany 1990 and 1994/95*

Fields of technology (Evaluation by enterprise management; several answers allowed)	Percentage of enterprises located in centres for innovation and establishment (n = 173)	
	1990 (%)	1994/95 (%)
Information and communication	32.0	33.5
General services	23.8	21.8
Measurement and regulation technology	20.3	22.5
Automation, robotisation	15.7	17.2
Mechanical engineering	14.5	12.2
Energy and environment technology	14.5	15.6
Chemistry technology	11.6	10.5
Production processing	9.9	10.3
Medical technology	8.1	7.2
Laser technology, optics	7.6	8.7
Materials technology	5.8	7.4
Biotechnology	5.2	5.8
Traffic and transport technology	5.2	6.2
Others	14.5	11.5

Source: Pett, A. Technologie- und Gründerzentren, Verlag Peter Lang, Frankfurt a.M. 1994, p. 222; Baranowski, G. / Groß, B. (ed.) Innovationszentren in Deutschland 1994/95, Weidler Buchverlag, Berlin 1994

Table 5.3 points out the fields of technology applied by enterprises which are located in selected German centres for new firm innovation. Although several answers were allowed, the evaluation of the enterprise management (1990) and the identification of the enterprises based on survey data (1994/95) conclude that the structure of the enterprises located in such centres has changed in favour of increased percentages of high-technology industries. Examples include materials technology, laser technology, energy and environment technology, robotisation, measurement and information technologies. The 'degree of innovation' was found in 1990 to be 'high' in 47.0% of the investigated companies and 'medium' in 34.9% (Pett, 1994).

Table 5.4 demonstrates examples of 'innovation poles' in Japan. They combine activities based on different resources and rates of technological change; they are effective as centres for the location of evolutionary processes (Schmidt, 1994; 1995). The data reveal that the impact of high-technology industries on the economic structure of the 'innovation poles' is higher in Japan than in Germany.

The innovative environment influences the structure of entrepreneurial networks in the considered countries. This statement also covers Switzerland and other European countries. An additional example is Singapore. According to Table 5.5 considerable numbers of national projects have a 'very strong impact' on the development of important new technologies. The projects are financed by the government, but they are implemented by co-operating manufacturing companies, including R&D laboratories, universities, technology centres and other institutions – all of them functioning as 'innovation poles' based, to a large extent, on entrepreneurial networks (National Science and Technology Board, 1991).

Table 5.4: *Fields of technology applied in selected 'innovation poles' in Japan 1993/95*

Fields of technology (Evaluation by institution management; several answers allowed)	Visited institutions ('innovation poles') (1993: n = 20, 1995: n = 15)	
	1993 (%)	1995 (%)
Information and communication	31.1	32.5
General services	23.8	25.4
Measurement and regulation technology	19.5	20.7
Automation, robotisation	19.3	20.5
Mechanical engineering	18.5	19.3
Energy and environment technology	12.2	11.8
Chemistry technology	11.8	12.4
Production processing	10.2	11.5
Medical technology	5.4	8.3
Laser technology, optics	9.2	9.9
Materials technology	8.7	9.4
Biotechnology	6.3	7.9
Traffic and transport technology	8.5	9.7
Others	11.6	10.8

Source: Own case studies

Table 5.5: *National projects of 'very strong impact' on new technologies in Singapore*

Important technologies \ National projects	Biomedical electronics	Personal Communication Network	Industrial Image Processors	Voice Recognition	National Home Care
Data Compression				x	x
Digital Signal Processor	x	x	x		x
Fuzzy Logic Systems				x	
Neural Networks				x	
Knowledge Engineering					x
Man-Machine Interface			x	x	
Networking, especially Multimedia		x	x		
Design technology		x			

Source: National Science & Technology Board, Republic of Singapore (ed.): Science and Technology. Window of Opportunities, National Technology Plan 1991, 1991, p. 115.

CONCLUSIONS

Starting from a theoretical hypothesis of network economics according to which the allocation of resources is optimised by flexible inputs of productive resources and entrepreneurial decision-making, this paper has attempted to point out the impact of entrepreneurial networks in an innovative environment. The preconditions of entrepreneurial networks were found to be determined by the behaviour of entrepreneurs, managers and employees under changing technologies and at increasing uncertainty. The patterns of entrepreneurial networks can be described by 4 types including subcontracting business, co-operative economic interfirm relations, human interaction based on non-economic relations and private and public or semi-public institutions. The subcontracting business and the institutions involved in the entrepreneurial networks were mainly investigated through reference to own case studies and surveys.

The results suggest first that innovative entrepreneurial networks are necessary in order to increase the competitiveness of enterprises under the conditions of accelerating technological change. Second, it is pointed out that innovative entrepreneurial networks need skilled, creative and flexible labour, risk-bearing capital, together with access to information on innovations and innovative management behaviour. In an innovative environment, entrepreneurial networks need creative and innovative employees to a growing extent.

REFERENCES

Acs, Z.J. and Audretsch, D.B (1992) *Innovation durch kleine Unternehmen*, Edition Sigma, Rainer Bohn, Berlin.

Baranowski, G. and Groß, B. (eds.) (1994) *Innovationszentren in Deutschland 1994/95, mit Firmenbeschreibungen*, Weidler, Berlin.
Dose, N. and Drexler, A. (eds.) (1988) *Technologieparks, Voraussetzungen, Bestandsaufnahme und Kritik*, Westdeutscher, Opladen.
Dosi, G. et al. (eds.) (1988) *Technical Change and Economic Theory*, Pinter, London.
Hofstetter von Trachselwald, St. (1990) Technologietransfer als Instrument zur Förderung von Innovationen in technologieorientierten, Klein- und Mittelunternehmungen, Thesis, St. Gallen.
Johannson, B., Karlsson, Ch. and Westin, L. (eds.) (1994) *Patterns of a Network Economy*, Springer, Heidelberg.
Lakshmanan, T.R. (1994) State Market Networks, in Japan: The Case of Industrial Policy in Johansson, B. et al. (eds.) *Patterns of a Network Economy*, Springer, Berlin.
National Science & Technology Board, Republic of Singapore (ed.) (1991) Science and Technology – 'Window of Opportunities', *National Technology Plan*, Singapore National Publishers, Singapore.
Nelson, R.R. and Winter, S.G. (1982) *An Evolutionary Theory of Economic Change*, Harvard University Press, Cambridge, MA.
Ogawa, E. (1984) *Modern Production Management. A Japanese Experience*, Asian Productivity Organisation, Tokyo.
Ogawa, E. (1993) Technology Transfer and Small Industry, in Pleitner, H.J. (ed.) *Small and Medium Sized Enterprises on Their Way into the Next*, Schweizerisches Institut für Gewerbliche Wirtschaft, St. Gallen.
Pett, A. (1994) *Technologie- und Gründerzentren, Europäische Hochschulschriften*, Reihe V, Bd. 158, Peter Lang, Frankfurt a.M.
Pohl, M. (ed.) (1995) *Japan 1994/95, Politik und Wirtschaft*, Institut für Asienkunde, Hamburg.
Schmalen, H. and Pechtl, H. (1992) *Technische Neuerungen in Kleinbetrieben*, Poeschel-Verlag, Stuttgart.
Schmidt, K.-H. (1984) *Die Arbeitsmotivation in Klein- und Mittelbetrieben Japans, in: Klein- und Mittelbetriebe – Motoren unserer Wirtschaft, Festgabe für E. Brugger*, Schweitzerische Volksbank, Bern.
Schmidt, K.-H. (1988) *Innovationen unter Unsicherheit. Rahmenbedingungen, Fallstudien und Strukturanalysen in Südafrika. Europäische Hochschulschriften*, Reihe V, Bd./Vol. 929, Peter Lang, Frankfurt a.M.
Schmidt, K.-H. (1989) *Strategien innovativer kleiner und mittlerer Unternehmen in der Schweiz*, IGW Impulse, 5, Schweizerisches Institut für Gewerbliche Wirtschaft, St. Gallen.
Schmidt, K.-H. (1990a) Innovations, uncertainty and size of the firm: a comparison between South Africa, Western Europe and Japan, *South African Journal for Entrepreneurship and Small Business*, 2, 1, April.
Schmidt, K.-H. (1990b) *International Comparison of Subcontracting Business in West European Countries and Japan. The Study of Business and Industry*, 7, edited by The Research Institute of Commerce, Tokyo.
Schmidt, K.-H. (1991) Cooperation of small and medium enterprises of economic integration, *South African Journal for Entrepreneurship and Small Business*, 3, 2, Dec.
Schmidt, K.-H. (1994) *Regional Innovation Poles in Japan*, University of Paderborn, Paderborn.
Schmidt, K.-H. (1995) Innovation poles: theoretical concepts and empirical data from Japan and Germany, *Industry & Higher Education*, 9, 1, February.
Schmidt, K.-H. (1996) Skill formation, labour mobility and network evolution – theoretical ideas, organisation and implementation, Proceedings of the 41st ICSB World Conference, Stockholm.
Schumpeter, J.A. (1942) *Capitalism, Socialism and Democracy*, Harper Brothers, New York.
Stoneman, P. (1987) *The Economic Analysis of Technology Policy*, Clarendon Press, Oxford.
The STRATOS Group. Bamberger, I. et al. (eds.) (1990) *Strategic Orientations of Small European Businesses*, Stratos, Aldershot.
Teubal, M. and Zuscovitch, E. (1994) Demand Revealing and Knowledge Differentiation Through Network, Evolution, in Johansson, B., Karlsson, Ch. and Westkin, L. (eds.) *Patterns of a Network Economy*, Springer, Berlin.
Witt, U. (ed.) (1990) *Studien zur Evolutorischen Ökonomik I, Schriften des Vereins für Socialpolitik*, Band 195/I, Duncker & Humblot, Berlin.

CHAPTER 6

Entrepreneurial Innovation Networks: Small Firms' Contribution to Collective Innovation Efforts

MANUEL LARANJA

INTRODUCTION

There have been a number of empirical studies (Oakey, 1984; Friar and Horwitch, 1986; Dodgson and Rothwell, 1989; Rothwell and Beesley, 1989; Dodgson, 1990; Dodgson and Rothwell, 1990) that emphasise the important complementary role played by external linkages in the innovation process of small firms. Dodgson and Rothwell (1989, p.148), in their study of technology-based firms in Europe, pointed out that:

> External technical expertise and internal knowledge generation were seen in all cases as complements in the overall process of technological know-how accumulation.

These linkages assume many different forms. While *formal* linkages such as joint ventures, subcontracting, licensing, co-production, joint- and cross-distribution, R&D collaboration, etc., are used to conduct research, product development, manufacturing and marketing, *informal* linkages are used for information gathering and uncertainty reduction, and they are often grounded in social and cultural environments. External formal linkages may also take the form of financial linkages or equity holdings, and often involve the participation of the same people in the administration boards of different companies, especially in the case of small start-up firms.

Moreover, external linkages between firms and universities and public institutes are not a new phenomenon (Rosenberg, 1976). According to Camagni (1991, p.137) it is important to recognise an:

> increasing or booming utilisation of those new forms of external development (i.e. universities) by firms of various sizes, particularly in those areas of production characterised by fast innovation and technological change like electronics equipment, telecommunications, semiconductors, software and factory automation devices : in a word, the information technology sector.

According to Freeman (1990) there is quantitative evidence of a strong upsurge of collaborations related to new technologies (mainly Information Technology). Also, there is evidence of a qualitative change in the nature of subcontracting and of collaboration with research institutes and government sponsored R&D programmes. Freeman (1990) also points out that small firms are by no means excluded from these changes, and that they simultaneously may be involved in more than one form of networking.

For the case of the small firm starting up there are few or no studies addressing the issues of networking. Elsewhere, this author has argued that developing, manufacturing and commercialising its products and services cannot be an isolated task for the small firm but rather a collaborative venture, centred upon a founding entrepreneur's personal contacts and involving different kinds of linkages with various organisations (Laranja, 1995). A useful means of considering linkages is to distinguish *upstream linkages* – such as contract R&D, technical information exchange, recruitment (which relate to the entrepreneurial task of acquiring and further developing or adapting technology) from *downstream linkages* – such as marketing and distribution agreements (related to the entrepreneurs' approach to market-introduction of relatively new product or services).

This paper focuses on particular cases of *upstream* and *downstream* linkages, which for the case of small start-up firms tend to be centred upon an entrepreneur's personal contacts (Roberts, 1991) and involve a spatially bounded social and informal dimension.

CONCEPTUAL APPROACHES TO NETWORKS

Conceptual approaches, such as *complementary assets*, *networks of innovators* and *innovative milieux* are particularly appropriate to our understanding of external linkages in start-up firms. The following section reviews some important aspects of these concepts.

Complementary assets in less favoured regions of Europe

Knowledge-intensive activities in small start-up businesses with no obvious economies of scale (or where economies of scale are external to the firms) (Teece, 1986), tempered by the independent drive and fears of losing management control on the part of founding entrepreneurs, are indicative of the conflict between a desire to remain small, on the one hand, and of the need to overcome the disadvantages of size by fostering intensive external interactions in various external domains, on the other hand.

The framework proposed by Teece (1986) suggests that firms seek collaborations with other organisations in order to access *complementary assets* needed for arriving at a complete final product, and its successful commercialisation. However, Teece's framework, for the case of small technology specialised start-up firms, must be further qualified.

First, in accessing sources of technology and technical information, small entrepreneurial firms may attempt to interact with local universities or public research institutes. The increasing volume of technological know-how needed for understanding and adapting technology originated abroad encourages the small firm to use universities as technology information sources and also repositories of high qualified engineering resources. Whilst this perspective is common in more advanced economies, for countries such as Portugal, enterprises in general, and small technology-based firms in particular, face greater difficulties in accessing R&D and technical information from the local universities.

Second, linkages in the form of commercial and marketing agreements are also an important component of the small firm's efforts to access complementary assets. The focusing of small entrepreneurial firms' resources upon specific technological sub-systems encourages the firm to establish marketing agreements with local sales subsidiaries, or with Value Added Resellers (VARs). After-sales services may also be a motivation for small firms to establish marketing agreements with other large or small firms. In Portugal, in the case of niche-software companies, it is often particularly advantageous for the small firm to engage in commercial agreements with local subsidiary distributors of hardware computer equipment who can market hardware and software together (Marques and Laranja, 1994).

Networks of innovators and the innovative milieux

Another helpful framework for networking in small 'start-up' firms is that proposed by Camagni (1991). He defines *innovation networks* as representing market-mediated, organisational and territorial *formal* relationships, while network relations of a mainly *tacit* and *informal* nature are referred to as *milieu* relationships (Aydalot and Keeble, 1988; Camagni, 1991; Castells and Hall, 1994). An *innovative milieu* is composed of a complex set of actors (firms, public administration, financial system, systems of R&D, system of services), operating in a particular territory, and the recognised relationships, mainly person- and socio-embodied, maintained between them. The concept of an innovative milieu therefore envelops the spatial characteristics of socio-economic, institutional, and cultural environments conducive to innovation. In other words, *innovative milieux* are necessarily spatially bounded, in the sense that they show some degree of cultural and social homogeneity across space.

For small technology-specialised small firms starting up in Portugal the concepts of networks of innovation, and informal innovative milieu can help to understand the firms' tendency to agglomerate around Lisbon and Oporto. Since these large urban areas are economically diversified, they offer greater possibilities for engaging in formal and informal tacit interactions.

Entrepreneurial innovation networks

There is a fundamental feature of external linkages and networks for firms starting up, which it is important to note. For small start-ups, networks are centred essentially upon their founding entrepreneurs' informal interactions with other individuals (Roberts, 1991), which perhaps is more strongly associated with location or spatial considerations. Birley (1985) claims that such networks of personal contacts are essential for credibility building, and for assembling the elements and resources needed for launching a firm. The same line of argument has also been put forward by Aldrich and Zimmer (1986), who point out the importance of informal and social networking for the successful launch of a new firm.

The understanding of the role that small start-up firms play in the context of Portugal requires further qualification of the concepts briefly outlined above. For the purpose of the following case studies we relied on a definition of *entrepreneurial innovation networks* as:

> Institutional and individual arrangements or networks corresponding to formal and personal informal linkages supporting technological innovation-imitation, through the encouragement of risk-taking and learning in small entrepreneurial firms.

Entrepreneurial innovation networks are built from formal collaboration agreements and are arguably supported, to a greater extent, by mainly person- and socio-embodied informal relations centred around the founding entrepreneurs. Formal and informal linkages in the network carry market and non-market mediated flows of goods, information and knowledge. These latter artefacts are particularly difficult to empirically capture accurately. Also, in these flows, there may be technological 'spill overs' and 'externalities' which are of fundamental importance for the emergence and early stage development of small technology-based firms.

In the following sections 2 cases of such networks are described. The approach is similar to that outlined by Håkansson (1987), but in this case applied to start-up firms focused upon information technology, located in Portugal. The gathering of information on events taking place in different locations was based upon direct in-depth interviews with founding entrepreneurs of small firms involved in the network and with managerial staff and engineers working at related companies and organisations.

THE DEVELOPMENT OF AN OFFICE SYSTEM IN AN ENTREPRENEURIAL NETWORK

An electronic office system comprises a computer running specialised office-application software such as spreadsheet, wordprocessor, communication interfaces for fax and e-mail, a database management system and specialised routines for processing documents, file handling and storage.

An extensive survey of activities undertaken by different actors showed that a small entrepreneurial start-up firm named SMD, founded in 1987, built up a network of linkages with other firms allowing further development and introduction in the market of 2 products: a software product for office automation, named ELENIX, and a hardware product for the front-office automation of the Post Office company.

These collaborations or linkages formed what we may call an *entrepreneurial innovation network* in which each organisation contributes by specialising on a particular function. Figure 6.1 illustrates the office technology network, by indicating the names of the most relevant organisations contributing to the final innovation outcome.

The university research institute (INESC – Institute of Engineering and Computer Systems) developed a first prototype (including both hardware and software) and played the part of the R&D unit to the overall network, allowing further co-development of the initial prototype. INESC's links to sources of technology abroad, through ESPRIT, were also found to be an important aspect of the contribution of university based R&D activities to the entrepreneurial network.

The small entrepreneurial firm (SMD) was dedicated to the commercialisation and further development of the original prototype developed by INESC. However, interestingly, the firm was founded by outside entrepreneurs who took INESC's development at an early stage, envisioning its further development into commercialisation. Basically, SMD performed a bridging role, positioning itself as the marketing and commercial 'front-end' of the network by interacting with customers and distributors. SMD correctly perceived the need to establish marketing distribution agreements for office application software, while it chose to follow a hardware 'niche' strategy taking advantage of on-going projects at INESC for the specific needs of the Post Offices.

Also playing an important role in the network were small-scale manufacturing firms for the hardware system which SMD developed for the Post Office front-office automation function. Subcontracting manufacturing enabled SMD to remain essentially a product development and marketing company, with no manufacturing capacity.

An important role in the office-technology entrepreneurial network is that played by local sales subsidiaries of hardware vendors who collaborate with SMD for the distribution and commercialisation of office application software. These collaborations allowed SMD to remain focused on the co-development of products with INESC, while taking advantage of the credibility and the marketing of complementary assets controlled by hardware sales subsidiaries such as Unisys, Digital, IBM, etc. These firms often choose to include in their final products tailored software which caters for language and specificities of local users.

Finally, other important actors in the entrepreneurial network are the final users of office technology. Public administration and, to a lesser extent, industry played an important role in providing demand opportunities which could be met by the small start-up firm. Moreover, the Post Offices not only provided a demand opportunity but also collaborated extensively in the testing of early prototypes, hence performing the role of 'early users'.

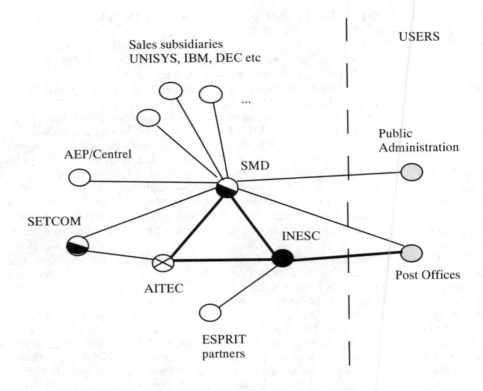

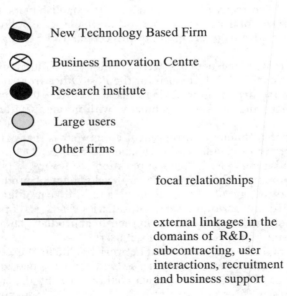

Figure 6.1
The office technology entrepreneurial network

AUTOMATIC TELLER MACHINES AND EFTPOS TERMINALS

A system that provides the bank customer with the possibility of on-line interactive services with its bank consists of a terminal for inputs, a means of identifying the customer, an electronic network to transmit data from the terminals to the data-management system at a central electronic-funds transfer clearing house, and a process to organise and transmit data gathered at the bank's existing batch processing systems. At the input end of these networks one can normally find different kinds of terminals, such as: ATM (Automatic Teller Machines) terminals, EFTPOS (Electronic Funds Transfer at the Point of Sales) terminals, Videotex and PCs.

In Portugal the establishment of the electronic data network supporting money transactions involved small entrepreneurial start-ups who collaborated extensively with local banks and universities, for the formation of the network and the manufacturing of terminal equipment.

The network arrangement that assisted the small start-up firms in their entrepreneurial task was found to consist of a chain of R&D, manufacturing and marketing interactions involving the local engineering polytechnic university (ISEL), SIBS, and various small technology-based firms as illustrated in Figure 6.2.

ISEL, the Engineering Polytechnic University of Lisbon, helped the electronic-money clearinghouse SIBS to develop specific terminal equipment to be used in banking networks. Such developments were apparently motivated by reasons related to security standards. However, economic factors were also relevant and the lower price of the terminals designed to speed adoption, particularly amongst the smaller retailers, also played an important role.

The banking electronic-money clearing house SIBS was the focal point of the whole entrepreneurial network. This company (initially owned by 11 local banks) played a central role of co-ordinating the entrepreneurial initiatives, including setting up the electronic money network and promoting collaboration between various other complementary organisations. Through the co-ordination of SIBS it was possible to build strong interactions among the R&D, manufacturing and marketing functions within the network.

Companies manufacturing the terminal equipment and collaborating to some extent in its development correspond to this paper's definition of small entrepreneurial start-up firms. Two of the companies (Gain and Papelaco) exploited small-batch manufacturing facilities and undertook the task of fabricating the hardware and final assembly. Another entrepreneurial small firm (Copinaque) was also involved in the network undertaking distribution of EFTPOS equipment amongst the retail sector.

Finally, a most important category to consider was the final users of both ATM and EFTPOS terminal equipment. These included the banks for ATM and large retailers (including petrol stations) or small outlet shops, for EFTPOS equipment. Indirectly, through their joint-venture SIBS, the banks were effectively creating a market opportunity for the small technology-based firms, and enhancing their technological capabilities by stimulating collaboration between the firms and a local university.

FINAL DISCUSSION

This final part of the paper brings together the findings, and shows how they contribute to an understanding of the patterns of external linkages which may be expected for small entrepreneur-owned start-ups.

First, although we cannot claim that universities in Portugal are extensively linked with small start-up firms, it appears that, in some cases, they may play an important role. In the face of scarce technological resources controlled by the small firms, the

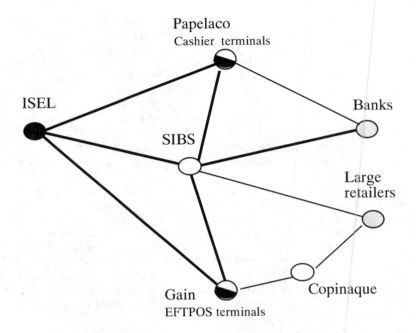
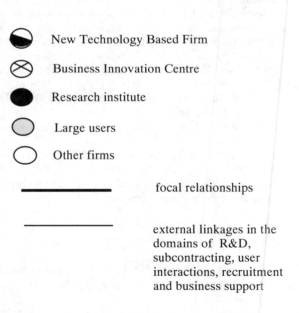

Figure 6.2
Entrepreneurial innovation network underlining the development, manufacturing and commercialisation of EFTPOS and ATM terminals

university effectively offers its R&D capabilities to the entrepreneurial network. Overall in the 2 cases, the university also helped to establish flows of technical information and enhanced the R&D skills of other actors in the network. If they participate in European R&D programmes, universities may also introduce and expose other members of the network to international technology trends and state-of-the-art, as it happened with the technology transfer between ESPRIT, INESC and SMD.

Second, the case studies suggest that an important source of linkage opportunities are local *large* users who engage in a process of technological change. We have seen in that the Post Offices, public administration and the banks have created important technological 'niche' opportunities that could be met by indigenous small technology intensive firms starting up. It should be noted that local users may provide more than just market opportunities since they provide more than a purchase relationship with the small firm. The user effectively may contribute to the entrepreneurial network by establishing collaboration links, envisioning testing and the further development of a particular product. Collaborative linkages with users, within the framework of entrepreneurial networks, are, according to the above evidence, motivated by technical reasons (i.e. by the need to adapt and localise products and concepts initially developed elsewhere, needing some modification in order to meet local requirements). Apparently, as argued by Maxwell (1983), 'ready made' solutions from abroad may not cater for particular requirements of local large users, hence the opening up of collaboration opportunities between the local technology-based start-ups and the local users.

This was the case of collaboration between the Post Offices, INESC and SMD, and it appears also to have been the case of collaboration between the banks (SIBS), ISEL and the small technology-based firms Gain and Papelaco. The Portuguese banks had their own security and user-interface requirements which could not be entirely satisfied by ready-available terminals from abroad.

However, it may be that factors other than technical differences arising from adaptation, such as after-sales services, user assistance and training, play a relatively more important role in influencing the user purchase decision.

Third, the small entrepreneurial firm plays the most important role in the entrepreneurial network. The small firm is often in charge of bridging between upstream sources of technological knowledge and commercialisation and establishing a direct interaction with the final user. To a lesser extent, small technology-based firms may also play a manufacturing role. Small firms may be dedicated to the exploitation and improvement of their production processes, while linking upstream with other small firms (and possibly universities for product development), and downstream with other small or large firms who perform the role of commercial distribution.

Finally, one must also point out that, while integration among R&D, manufacturing, marketing and sales has been noticed as a fundamental characteristic of technological innovation in the context of a firm as the single innovator (Imai et al., 1988; Rothwell, 1990), the cases presented here appear to suggest that, in the context of a group of firms and institutions, there must also be a good co-ordination and integration of several functions. Clearly for less advanced countries such as Portugal innovation studies should take into account that integration between R&D, marketing and sales is often undertaken within the context of a complex network of formal and informal interactions involving small technology-based firms, local universities, local sales subsidiaries of multinational companies and value added resellers.

REFERENCES

Aldrich, H. and Zimmer, C. (1986) Entrepreneurship through Social Networks, in Sexton, D. and Smilor, R. (eds.) *The Art and Science of Entrepreneurship*, Ballinger, Cambridge, Mass.

Aydalot, P. and Keeble, D. (1988) Technological Trajectories and Regional Innovation in Europe, in Aydalot, P. and Keeble, D. (eds.) *Technology Industry and Innovative Environments: The European experience*, GREMI–Groupe de Recherche Européen sur les Milieux Innovateurs, Routledge, London.

Birley, S. (1985) The Role of Networks in the Entrepreneurial Process, *Journal of Business Venturing*, 1, 1.

Camagni, R. (1991) Local milieu, uncertainty and innovation networks: towards a new dynamic theory of economic space, in Camagni, R. (ed.) *Innovation networks: spacial perspectives*, GREMI – Groupe de Recherche Européen sur les Milieux Innovateurs, Belhaven Press, London and New York.

Castells, M. and Hall, P. (1994) *Technopoles of the World: The Making of Twenty-First-Century Industrial Complexes*, Routledge, London and New York.

Dodgson, M. (1990) Technological Learning, Technology Strategy and Competitive pressures, presented at the Workshop Process of Knowledge Accumulation and the Formulation of Technology Strategy, May 20–23, Kalundborg, Denmark.

Dodgson, M. and Rothwell, R. (1989) Technology strategies in small and medium-sized firms, in Dodgson, M. (ed.) *Technology Strategy and the Firm: Management and Public Policy*, a SPRU publication, Longman, London.

Dodgson, M. and Rothwell, R. (1990) Strategies for Technological Accumulation in Innovative Small and Medium-Sized Firms, paper presented at the Symposium on Growth and Development of Small High-Tech Businesses, 2–3 April, Cranfield Institute of Technology, Cranfield.

Freeman, C. (1990) Networks of Innovators: A Synthesis of Research Issues, paper presented at the International Workshop on Networks of Innovators, May 1–3, Montreal, Canada.

Friar, J. and Horwitch, M. (1986) The emergence of technology strategy: a new dimension of strategic management, in Horwitch (ed.) *Technology in the Modern Corporation: a strategic perspective*, Pergamon, Oxford.

Håkansson, H. (ed.) (1987) *Industrial Technological Development: A Network Approach*, Croom Helm, London.

Imai, Ken-ichi, Nonaka, I. and Takeuchi, H. (1988) Managing the New Product Development Process: How Japanese Companies Learn and Unlearn, in Tushman, M. and More, W. (eds.) *Readings in the Management of Innovation*, (2nd Ed.), Ballinger, Cambridge, MA.

Laranja, M. (1995) Small firm entrepreneurial innovation: the case of electronic and information technologies in Portugal, DPhil Thesis, University of Sussex Science Policy Research Unit.

Marques, J. and Laranja, M. (1994) *As Tecnologias de Informação em Portugal: Importância, Realidade e Perspectivas*, DGI Ministry of Industry and Energy, Lisbon.

Maxwell, P. (1983) Specialisation decisions in electronic production – lessons from the experience of two Argentine firms, in Jacobson and Sigurdson (eds.) *Technological Trends and Challenges in Electronics: Dominance of the Industrialised World and Responses from the 3rd World*, Research Policy Institute, University of Lund, Sweden.

Oakey, R. (1984) Innovation and Regional Growth in Small High Technology Firms: Evidence from Britain and the USA, *Regional Studies*, 18, 3, pp. 237–51.

Roberts, E. (1991) *Entrepreneurs in High-Technology: lessons from MIT and beyond*, Oxford University Press, New York.

Rosenberg, N. (1976) *Perspectives on Technology*, Cambridge University Press, Cambridge.

Rothwell, R. (1990) External Networking and Innovation in Small and Medium-Sized Manufacturing Firms in Europe, presented at the conference Network of Innovators Workshop, May 1–3, Montreal, Canada.

Rothwell, R. and Beesley, M. (1989) The Importance of Technology Transfer, in Barber, J., Metcalfe, S. and Porteous, M. (eds.) *Barriers to Growth in Small Firms*, Routledge, London.

Teece, D. (1986) Profiting from technological innovation: implications for integration, collaboration, licensing and public policy, *Research Policy*, 15, pp. 285–305.

CHAPTER 7

Support of Technology-Based SMEs: An Analysis of the Owner Manager's Attitude

MAGNUS KLOFSTEN AND ANN-SOFI MIKAELSSON

INTRODUCTION

During the past few decades it has been shown that technological innovation has a key role to play in revitalising economically deprived business sectors and localities (Westhead and Storey, 1994). Therefore governments in European and North American countries are increasingly concerned with supporting the creation and growth of technology-based firms (Oakey, 1985; 1991). An important part of these initiatives is the support of the development of efficient support programmes to assist firms in the acquisition of knowledge and skills, essential for a successful business (Kirby, 1990; Jones-Evans and Kirby, 1995; Klofsten, 1995).

Studies have shown that there are a number of factors which are important when developing and managing successful support programmes for small businesses. In 2 case studies of the support of local high-technology firm clusters, at least 5 such factors were found (Klofsten and Jones-Evans, 1996; Autio and Klofsten, 1996). These studies strongly suggest that greater attention should be placed on the real needs of member firms. Other important factors determining success were the existence of a core group of people, a clear activity focus, credibility, and close links with a university.

Delatte and Baytos (1993) give a number of guidelines for managing training programmes such as the need to distinguish between education and training, and the incorporation of both into a programme. Training should also be positioned as a part of an overall strategy for managing diversity of need. Such provision should be based on a thorough needs analysis, including diverse input into the design process to increase specific relevance. Any programme should be thoroughly tested before introduction to reduce risk and generate enthusiasm. These should be of a mixture of internal and external firm resources to enhance efficiency and credibility, and incorporate diversity of education and training into a core curriculum so that it becomes an organisational on-going way of life, and not a one-shot programme. Similar factors are articulated in Johnson (1987) where he studied the 'Training and Enterprise Programmes' and found that key factors in training firms were stimulating participation, encouraging identification and the articulation of personal learning objectives, giving examples to which the firms can relate, using a variety of learning/training methods, being open to all questions, and presenting first-hand experience.

The above success factors or guidelines suggest how ideal support programmes should be managed. An important point is how the actual business support is delivered to the firms and how they experience the actual support – or in other words how effectively the available resources supplied by the supporting organisations are matched to the demands of businesses for different forms of supplementary support. Gibb (1992), Kirby (1990) and Mönsted (1986) are of the opinion that a gap between supply and demand exists as a result of the fact that the support that is given, often is not suited to businesses in general, or small firms in particular. The authors address a number of critical points concerning barriers to reaching the small firm (i.e. the demand side) and problems in styling the support for the small firm (i.e. the supply side).

Barriers on the demand side could be described as the firm's scepticism concerning the value of support, their inability to pay, their lack of time, and the reality that they prefer investing money in activities that seem to provide a direct return in terms of output rather than indirect activities such as training. Examples of barriers to the supply side could be that supporting institutions employ experts who do not have experience of small businesses, they emphasise theory and concept when the owner-manager is more concerned with practical issues, and that much of the support is delivered with the trainer dominating the classroom and the audience in a passive role, except when questions are invited. This style of learning may be inappropriate to those who are more comfortable with 'learning by doing'.

AIM AND SCOPE

The above statements about barriers to successful training are based on experiences of various support programmes. However, to our knowledge, they have not been tested empirically on a group of firms in general and technology-based firms. This paper studies how the firms (the entrepreneurs) perceive the offered support, partly in terms of the utility of the resources supplied in support programmes, and partly in terms of the firms' attitude towards the ability of the supply side to satisfy their real needs. The research questions addressed are as follows:

- *Do any barriers exist in terms of the demand side concerning the firms' actual situation and the perceived benefit of taking part in different support programmes?*
- *Are there any barriers on the supply side concerning the firms' view of the support organisations' ability satisfactorily to develop programmes?*

The purpose of this paper is to study whether a gap between supply and demand, as described above, exists, and to explore how this possible gap can be bridged to create efficient business support programmes. When defining support, we have used the definition from Klofsten (1995) where support constitutes measures for increasing the competence level of firms. These measures have the following goals: to increase the number of business start-ups, to improve the quality of these firms, to increase the chances of survival of new and existing firms, and to encourage their growth and development.

METHOD AND DATA

The firms in this study have been selected from a database, developed at Linköping University, consisting of 158 technology-based firms located in the vicinity of Linköping. From this sample, 59 firms were excluded for the following reasons: the firms were dormant or no longer in existence; they had changed names and were therefore represented 2 or 3 times on the list; the firms were too large (>250 employees, e.g. Ericsson Radio Systems or Saab Scania Aircraft Division) to be considered as small firms, or they were a subsidiary of a large company.

After excluding the above firms, a questionnaire was sent to the chief executive officer, or to a staff member, of the remaining 99 firms, from which 62 replies were obtained (i.e. a response rate of 62%). The questionnaire included 42 questions where the majority were closed alternatives representing 80 coded variables. The respondent was informed of the purpose of the questionnaire by a letter of introduction in which types of business support were defined. The questionnaire was divided into 4 main sections as follows:

- The firm's background and general characteristics including: product, service, year founded, turnover, number of employees, ownership, level of prosperity (i.e. growing, maturing or declining), stage in the lifecycle (start-up, mature, decline), intentions for the business (expansion, stagnation, decline), board of directors, customers, markets, education, previous experience, and if the respondent had attended a business support programme earlier. The final background question dealt with the firm's resources and external relationships where a rating (Likert scale 1–5) of the actual availability of both internal and external resources to the firm, and the importance of having those available, was used.
- The attitude of the firm to support programmes was also tested. The respondent was asked to answer 11 questions on the same type of Likert scale as above on the amount of the firm's time available, when it comes to participation in a support programme and whether this is insufficient – or sufficient? They were also asked if they preferred learning from written material to learning from real life experiences.
- The firm's attitude towards actors who arrange support programmes and the content of the programmes including whether the support organisations employ experts or non-experts to deliver their expertise to participants and whether training focused mostly on theory rather than on practical knowledge.
- The firm's view on what actions ought to be taken to bridge the gap that might exist between supply and demand of business support including specific suggestions on which areas of competence to cover, how to organise programmes, the amount of time to be spent, and the cost of participating in support programmes.

The study has a number of limitations. The main source of information has been the questionnaire which was answered by the chief executive officer or member of staff although some independent firm documentation was available. As shown in Arbaugh and Sexton (1994) this can limit the contribution of entrepreneurial teams to the management process of the organisation and this factor (e.g. other significant information about business development both written and oral) should be taken into account in more detailed follow-up studies to research. The survey questionnaire process depends greatly on subjective memories, evaluations of past events and personal viewpoints and interpretations of the firm which shed doubt on the validity and reliability of such evidence. Unfortunately, comprehensively validating data through other sources of information would, in this case, have been too costly. It is important to keep these factors in mind while examining the results of the study. It should also be stated that the analysis below shows that a number of the firms do not have a clear opinion about some of the issues described above which include taking part in a business support programme. Other respondents have not been affected by the issues, and have therefore chosen not to adopt an opinion.

ANALYSIS AND RESULTS

After a short note on sample characteristics, results were analysed concerning the firms' (entrepreneurs') actual situation, the perceived benefit of taking part in different support programmes, and their view of the support organisation's ability satisfactorily to develop programmes. We have chosen variables which are clearly negative or positive in that they are clearly located on one or the other side of the 'neutral' line 3 on a 5 point Likert scale.

Sample characteristics

The firms in the study represent a number of high-technology areas such as software development, sensor technology, image processing, fibre composites, electronics and vacuum technology where the majority of the firms (i.e. 87%) offer customers a combination of product and service or service only. Other sample characteristics concerning age and size are:

- Age of the firms (yrs): mean (8), median (7)
- No. of empl.: mean (26), median (5)
- Range (no. of empl.): 1–89
- Turnover (SEK million): mean (4), median (24)
- Range (SEK million): 0.5–80

Concerning the stage of development, 23% of the firms were in a start-up or early development stage, the majority (i.e. 52%), in a growth stage, while 25% were in the mature or declining stage. In 79% of the cases, the firm reported an intended expansion in the future. When it comes to the availability of internal and external resources, and the importance of having these available, it is clear that the firms lack both marketing competence and personnel. Not surprisingly the firms judge themselves strong on the technical side. A majority of 67% of the firms had not taken part in a business support programme.

Participation in support programmes

Of the 11 variables, 8 were valid and are listed in Table 7.1. When it comes to taking part in support programmes, the firms judge the time available and financial resources to be either insufficient or very insufficient. The same attitude applies to their resources for implementing what has been learnt on the programmes: once back in their firms they often lack the time and resources to implement the acquired competence. They have a positive or very positive attitude to conventional support programmes but, at the same time, disagree or strongly disagree to learning from written material (articles, books, cases etc.). They prefer learning from their own and others' experiences.

The firms agree or strongly agree on investing money in activities which provide a direct return in terms of output and fulfilling customers' needs rather than indirect activities such as support programmes. They agree or strongly agree about their unawareness of support programmes on offer, resulting from limited capability to scan the environment. It is disagreed or strongly disagreed that the firms prefer the trust of peers and contacts rather than taking part in business support programmes.

Attitude towards support organisations

The analysis showed that 5 of the 10 variables were valid (see Table 7.2). Concerning the firms' attitude to the support organisations' ability to deliver support properly they agree or strongly agree that these organisations employ experts with experiences mainly from large organisations and thereby have little knowledge of the small business.

In all other valid cases, the firms disagree or strongly disagree. This means that the negative statements, expressed in aim and scope, concerning inflexible timetables

Table 7.1: *Participation in supporting programmes – valid variables*

Question	Opinion	No. of Cases
Time spent participating in support programmes	Insufficient to very insufficient	59
Financial resources, concerning participation in support programmes	Insufficient to very insufficient	60
Resources to implement and conduct activities born out of a support programme	Insufficient to very insufficient	60
Attitude towards conventional support programmes	Positive to very positive	59
Prefer learning from written material to learning from experiences	Disagree to strongly disagree	59
Inclined to invest in customer visits with direct pay-off, rather than activities that might give indirect pay-off e.g. support programmes	Agree to strongly agree	60
Unaware of the training on offer and have limited capability to scan the environment	Agree to strongly agree	59
Prefer contacts and advice from our peers instead of taking part in support programmes	Disagree to strongly disagree	60

Table 7.2: *Attitude towards support organisations – valid variables*

Question	Opinion	No. of Cases
Specialists often have experience from large companies only	Agree to strongly agree	58
Training is conducted conventionally with a rigid and inflexible timetable	Disagree to strongly disagree	56
Training institutions ought to engage trainers with a high theoretical knowledge instead of those with knowledge based on experiences	Disagree to strongly disagree	58
Material is often written in a language not suitable to the owner-manager	Disagree to strongly disagree	58
Training is based on one-way communication and does not stimulate learning by doing	Disagree to strongly disagree	58

controlled by the programme manager, theoretical emphasis in the programme, formal language in the material offered in programmes, and the one-way communication style in the support programmes have not, in the whole group of firms, been proven to be correct.

Comparative analysis

A comparative analysis was made dealing with different groups of firms in our sample and their attitudes towards business support programmes. There were a number of significant differences (at the 5% level) concerning variables such as the perceived lack of resources, actual and intended business expansion, the degree of internationalisation, management experience, the specific time when the firms were founded, and differences between firms who had already taken part or not taken part in a business support programme.

Access to different resources affects, to some degree, the attitude of the firms towards business support programmes. This is significant for those who are lacking resources such as production equipment, finance, marketing and personnel resources. Firms that assess themselves to lack production equipment, and are in an early development phase, believe that support organisations use specialists ($p=0.039$). In the cases where capital is lacking it is believed that the programmes have an inflexible timetable ($p=0.017$) and are conducted by programme managers with theoretical competence ($p=0.048$). The same results were achieved by Kirby (1990) where he found that there is a shortage of quality trainers who combine experience of the sector and training requirements of the firm with the skills in delivering training through the process of learning by doing. Kirby also states that training packages need to be more flexible than previously, and tailored more closely to the specific needs of the firms.

Firms lacking marketing resources feel that the supply of qualified trainers is scarce ($p=0.010$); this is also felt by those who lack personnel ($p=0.041$) and they prefer advice from peers ($p=0.05$). It appears to be the case that firms who argue that they lack some essential resources, and are thereby somewhat less well developed, are critical of formalised programmes, of trainers with a theoretical approach, consider supply of qualified trainers to be scarce and prefer to use their social network for consultation and to acquire whatever knowledge they feel is necessary. This finding is supported by Birley (1985), who found that firms initially preferred to rely on a 'so called' informal network, friends, family and relatives.

Firms who claim that they have sufficient skills in personnel management also consider themselves to have the financial resources to take part in support programmes ($p=0.012$). It seems to be the case that firms who have gained good skills in personnel management believe that it is important to invest financial resources in competence increasing programmes, and that these firms are more inclined to send their staff to business support programmes. The study by Kirby (1990) also indicates that managers who have received training are more likely to train their own staff. This category of firms believe that the supply of qualified trainers is sufficient ($p=0.000$). An interesting note is whether the interest in personnel issues also provides insights into organisations where good concepts and trainers are available.

Firms with experienced managements believe themselves to have less time to take part in programmes ($p=0.050$). It is possible to view this as management feeling that their skills were complete, with no need for extra training inputs from a business support programme. The consequences of this could be, as Kirby (1990) states, that many firms are simply unaware of their own management deficiencies and the training that is required. For example, another result suggests that experienced management believe training is, perhaps surprisingly, best carried out through one-way communication ($p=0.038$). However, since more interaction is clearly preferable, this result implies that, at least where more experienced management is concerned, there will be a greater need for interactive programmes.

The larger group of firms in the population employing less than 25 workers considered the form in which support programmes are conducted to be positive ($p=0.045$). The same category do not consider themselves to lack financial resources to participate in a support programme ($p=0.042$). This group is, overall, more positive towards business support. It appears to be harder to conduct programmes with the smaller firms since they probably have the most informal structure and process. In other words, it is difficult to reach the informal firms with programmes that are perceived as formal. This is also the case with firms intending to expand. These firms give priority to market and customer activities rather than participation in support programmes ($p=0.027$).

The degree of internationalisation seems to affect the opinions of the firms regarding support programmes. Firms with a relatively low degree of internationalisation have generally a negative attitude, especially concerning the lack of holistic view (p=0.003) and degree of formalisation (p=0.047). There is also a tendency that firms with a relatively high degree of international customers view themselves as having better financial resources to participate in support programmes (p=0.05).

The firms differ regarding start-up time, especially those begun during the period of prosperity, 1986–1990. Businesses started during this period tend to consider themselves to have resources to participate in various business support programmes (p=0.012). It is likely that it is easier to reach firms during times of prosperity in the sense that they have relatively better financial resources to participate in competence increasing programmes.

The question of what the firm thought of the financial resources available for participating in a support programme showed that those who had participated considered their resources as sufficient (or very sufficient) (p=0.020), as opposed to those who had not participated (p=0.020). This could mean that those who had participated had good experiences and thought that the value/gain had been proportional to the cost. The survey findings of Kirby (1990) show that those who have experienced management training were all very positive. It could also mean, however, that they had a different priority system and considered learning worthy of investment both in terms of money and time, while the others had different priorities and values.

Regarding the question dealing with the firm's view of support programmes, it is shown that those firms who had already taken part in programmes were far more negative to conventional and formalised programmes than their counterparts (p=0.028). The causes of their clear standpoint cannot be explained on the basis of the questionnaire, but a possible reason might be that they have experiences from taking part in either good or bad programmes, or both, and therefore know what design they prefer.

Suggestions for improvements to programmes

At the end of the questionnaire, the respondents were given space to comment on the content and organisation of business support programmes. Those who had previously taken part in a programme gave a number of reasons for choosing that specific programme. One particular reason which was given by a number of firms was that the programme had a good reputation in the eyes of other firm management, and that they had been encouraged to participate. The positive aspects of a successful programme were pointed out as being the opportunity to deeply discuss their own problems and exchange experiences with other firms, that the timing of the programme was precise and matched their needs of support at that time, and that the programme managers had practical experience of small businesses. Suggestions for programme topics varied but those often mentioned were marketing, managing growth, quality assurance, personnel development and internationalisation.

The suggestions given on how to better organise a programme centred mainly around 2 areas; first on how information on various forms of business support can reach the firm and, second, on the need to focus on the firm's individual customised programme needs. In the first case, it was suggested that small businesses should be connected to the Internet to create consistent networking interaction between firms and support organisations, as exemplified through the distribution of news letters and personnel visits. To better suit the content of the support programme to the firm's needs respondent firms considered that a thorough needs analysis of the participating firms ought to be conducted prior to the programme. Some firm managements also considered that business support services could be tailored to different categories of firms by offering customised programmes, either to technology-based firms, or programmes oriented to different stages of firms' development.

Willingness to pay for participating in a business support programme varied greatly between the firms, in that anything from between 2,000 and 50,000 Swedish Kronor was mentioned. However, the figure most mentioned was 10,000 Swedish Kronor. The majority of the respondents were opposed to differentiated fees based on the firm's ability to pay for participation in the same programme. Preferably, it was considered that the cost should be low enough for all firms with needs to be able to participate, although programmes aimed towards start-up firms, it was considered, should have a lower fee than those aimed at more established firms.

CONCLUSIONS

This paper has focused on the attitudes of firms towards taking part in support programmes. The attitude of the firm towards support programmes, and how any friction between demand and supply occurs, was also studied, as well as how this could be eased to achieve efficient support activities. A number of questions were posed from statements made by Gibb (1992) concerning the difficulties of reaching the firm both on the demand and supply side. These questions were tested quantitatively upon a group of technology-based firms.

This study shows that there exists to some extent a gap between the supply of and demand for business programmes. The gap is evident when the firms' participation in various programmes is examined. The firms (or their entrepreneurs) consider that they lack time and resources with which to participate, while also, if they were to participate, they would not have enough resources to implement what they had learnt from the programme. Other conclusions are that firms give priority to activities with direct pay-off such as customer visits, instead of participation in a business support programme where both the direct and future pay-off is uncertain. Firm executives are also unaware of the market supply of programmes, and consider themselves to have limited possibilities of finding them out. However, this gap is not as apparent in all cases. The firms, as a group, are quite positive towards participating in conventional programmes and they do not prefer advice from peers to the advice obtainable from programme managers.

Concerning attitudes towards support organisations, a problem was (for the whole surveyed population of firms) found in relation to only one of the variables. This was that support organisations were deemed to be often orientated toward large firms, and therefore tend to lack knowledge of the specific problems of small businesses.

The comparative analysis shows some interesting differences between various groups of firms, in which young and small firms were noted to be not a homogeneous group concerning their attitudes toward business support programmes. Firms that are less developed, and suffer lack of resources, are more negative to a number of factors including the use of specialists, a theoretical approach, inflexible schedules and formalised programmes, when compared to others in our survey population. These firms prefer advice from peers and friends rather than participation in business support programmes.

Firms who have increased their size, and reached a certain notable level of internationalisation, are overall more positive toward support programmes, even though they have limited time for participation. These firms demand that programmes have a holistic view, are more formalised, and have an interactive character. In firms with a relatively well developed personnel department, participation in support programmes is considered important and they allocate financial resources for this cause. It appears to be the case that these firms are more market oriented concerning the supply of various business support programmes.

Why do firms choose to attend specific programmes, and what is the success formula for conducting such programmes? It appears to be very important that the support organisation has high credibility. It is significant to note that there exists a word-of-mouth effect in that firms discuss support programme quality, and that a good reputation considerably facilitates the marketing of support programmes to the small business market. A successful programme is distinguished by the firm sensing that its particular problem has been defined and dealt with in the programme. Before and during a programme it is important to keep in close contact with the firms through, for example, news letters and personal visits.

IMPLICATIONS AND FUTURE RESEARCH

This study has several implications which can be summarised as follows. Firstly support programmes should focus on defining the firms' real needs in order to be able to deliver proper solutions. Furthermore it is important to stress the significance of demand pull in support programme delivery since supply push is all too often supported from public funds (1996).

It is important to be aware of the fact that small businesses are not a homogeneous population and that attitudes differ widely between single firms and groups of firms concerning the attitudes towards business support. The consequence of this is that it will be necessary to segment the small business support market, either in terms of business category or stage of development. It is also important to facilitate communication and networking between firms and support organisations. This means that the firms will have a better view of the offered programmes, and that support organisations will receive a better insight and understanding of small business problems. Finally the supply of good programme managers must be secured as consultants in support programmes. These implications support what has been articulated in other studies during the last few years such as:

- Caird (1992): There is a need for better collaboration between firms, universities and governments and more local practical assistance for innovative development within firms.
- Gibb (1992): The need for an 'entrepreneurial approach' in small business training.
- Klofsten (1995): Important success factors behind effective business support are an 'ability to meet real needs', a 'core group', a 'clear focus', 'credibility', and 'close relations between the support organisation and a university'.
- Ronstadt and Paulin (1995): Networking through 'CONNECT' to foster technology transfer and entrepreneurship.
- Storey (1994): Increased emphasis on 'selectivity/targeting' and 'special groups' in business policy is needed.

An important area for future study is how to develop business support programmes to suit the real needs of those firms that have growth potential, but lack some essential resources, and most importantly, have a negative attitude towards these programmes. Another interesting area for future research is to study the mechanisms behind a firm's real need for stimulation. How can the firm's real needs be determined? Is it possible to measure whether a firm is an effective recipient of stimulation? Which type of stimulation programmes are necessary? How can the type of stimulation to be used in a successful programme be determined?

A study could also be conducted of support organisations' attitude towards the demand side of business support. The same type of research questions as above could be asked such as: do any barriers exist on the supply side concerning the support organisations' actual situation and the perceived efficiency of managing support programmes, and are there any barriers on the demand side concerning the support organisations' view of their ability to develop satisfactory programmes?

REFERENCES

Arbaugh, J.B. and Sexton, D.L. (1994) Determining Planning Process Differences for Growth-Oriented New Ventures: A Qualitative Analysis of High-Tech and Non High-Tech Firms, *Journal of Enterprising Culture*, 2, 4, pp. 887–906.

Autio, E. and Klofsten, M. (1996) Local Support for Technology-based SMEs: Two Scandinavian Cases, paper for the RISE 96: Research on Innovative Strategies and Entrepreneurship, University of Jyväskylä, Finland, 17th–19th June.

Birley, S. (1985) The Role of Networks in the Entrepreneurial Process, Frontiers of Entrepreneurship Research, Babson Entrepreneurship Research Conference, Babson College Wellesley, MA, pp. 325–37.

Caird, S. (1992) What Support is Needed by Innovative Small Business?, *Journal of General Management*, 18, 2, pp. 45–68.

Delatte, A.P. and Baytos, L. (1993) 8 Guidelines for Successful Diversity Training, *Training*, January, pp. 55–60.

Gibb, A.A. (1992) Design Effective Programmes for Encouraging the Small Business Start-up Process, *Journal of European Industrial Training*, 14, 1, pp. 17–25.

Johnson, R. (1987) The Business of Helping the Entrepreneur, *Personnel Management*, March, pp. 38–41.

Jones-Evans, D. and Kirby, D.A. (1995) Small Technical Consultancies and Their Client Customers: An Analysis in North East England, *Entrepreneurship and Regional Development*, 7, pp. 21–40.

Kirby, D.A. (1990) Management Education and Small Business Development: An Exploratory Study of Small Firms in the UK, *Journal of Small Business Management*, October, pp. 78–87.

Klofsten, M. (1995) Stimulating the Growth and Development of Small Technology-Based Firms: An Entrepreneurship Model that Works, paper presented at the 15th annual Entrepreneurship Research Conference (The Babson Entrepreneurship Research Conference) at London Business School, 9th–13th April.

Klofsten, M. and Jones-Evans, D. (1996) Stimulation of Technology-Based Small Firms: A Case Study of University Industry Co-operation, *Technovation*, 16, 4, pp. 187–93.

Mönsted, M. (1986) Rådgivningsstrukturen og den lille virksomhed, in Bohman, H. and Pousette, K. (eds.) *Konferendokumentation småföretags-forskning i tiden*, pp. 437–58, Department of Management and Economics, Umeå.

Oakey, R. (1985) British University Science Parks and High Technology Small Firms: A Comment on the Potential for Sustained Industrial Growth, *International Small Business Journal*, 4, pp. 58–67.

Oakey, R. (1991) High Technology Small Firms: Their Potential for Rapid Industrial Growth, *International Small Business Journal*, 9, pp. 30–42.

Ronstadt, R. and Paulin, W. (1995) Influencing Entrepreneurship in the Technopolis – Can CONNECT of San Diego Be Replicated Elsewhere?, paper presented at the 15th annual Entrepreneurship Research Conference (The Babson Entrepreneurship Research Conference) at London Business School, 9th–13th April.

Storey, D.J. (1994) *Understanding the Small Business Sector*, Routledge, London.

Westhead, P. and Storey, D.J. (1994) An Assessment of Firms Located On and Off Science Parks in the United Kingdom, SME Centre, University of Warwick, Coventry.

PART V Trust in Forming and Operating Innovative IORs

CHAPTER 8

The Role and Means of Trust Creation in Partnership Formation between Small and Large Technology Firms: A Preliminary Study of how Small Firms Attempt to Create Trust in their Potential Partners

KIRSIMARJA BLOMQVIST

INTRODUCTION

Globalisation, rapid technological change and an intense race against time have forced both small and large high technology firms to consider co-operation as an alternative competition strategy. The idea of large and small high technology based firms joining forces to achieve a competitive edge by sharing know-how and resources in research, product development, production and marketing is widely accepted among practitioners and researchers (Doz, 1988; Forrest, 1990; Forrest and Martin, 1992). Contractor and Lorange (1988) even refer to 'eclectic atmospheres' bringing out innovations not likely to be achieved in any one parent organisation's mono-culture context. Segers (1992) refers to Rothwell, who states that the main advantages of small firms are 'people embodied', while those of large firms are predominantly 'resource embodied'. At best, both partners gain from collaboration: the large firm has financial resources, established distribution systems and marketing management know-how, while the small firm has research capabilities and innovative products to complement the large partner's breadth of research, product development, and consequent product range (Doz, 1988; Forrest, 1990; Segers, 1992).

The point of departure for this paper is an examination of the small high technology based firm that tries to internationalise its operations and, to do so, aims at establishing a partnership with a large potential partner. Small technology based firms often focus on 'niche' products and additional markets are often needed at a fairly early stage of the product life cycle. Especially in small countries, where domestic markets are also small, such firms are forced to internationalise early. The earlier and the larger the market share firms are able to gain, the better they are able to recoup their invested R&D expenses. However, in those industries where critical mass of finances and marketing resources required for the breakthrough is sizable, strong large firm partners may be the only solution.

Trust has been identified as a key factor in relationship development (Morgan and Hunt, 1994) and also in technology partnership establishment and management (Forrest, 1990; Håkanson, 1993). This preliminary study particularly focuses on the role of trust and the means of trust creation that small high technology firms use when attempting to form partnerships with larger firms.

PROBLEM SETTING AND RELEVANCE OF THE STUDY

In an initial phase, when a partnership is not yet established, and the decision whether to cooperate or not has yet to be made, the ability to trust a potential partner seems critical. Especially when establishing technology partnerships with small (often also young) firms, there are usually very few concrete matters to assess. Then the small technology based firm's ability to signal its competence and trustworthiness to a potential larger partner becomes vital (Meldrum, 1995). The question whether the small firm can trust the large firm's good intentions in a given situation, where the large firm is often the more powerful actor, is also a very important issue. However, this paper focuses only on the role of trust, and the measures small high technology firms take when attempting to create trust.

The present knowledge of how to build trust is weak. Trust is measured mainly on scales of how much, if any, the respondent trusts the other party. At its simplest form, the respondent is asked whether he or she trusts the other party. This does not, however, increase our understanding of what trust really is. The respondents might easily attach various meanings and contexts to trust, and thus not provide a consistent answer. Therefore a different approach to the study of trust is needed. Also Young et al. (1992) criticise the existing measures of trust by noting that they often only include single items, which tend to capture only a small portion of the construct. It could be easily argued that previous attempts to measure trust are not able to capture the whole phenomenon; trust being both context and situation-specific and perceived subjectively by each individual. Lewis and Weigert (1985) also argue that present trust measuring methodologies may easily have reductionistic consequences.

Zucker (1986) believes, 'that trust is routinely produced and that such production is fundamental to understanding any exchanges in a social system' (p. 59). He also notes that trust is often defined in terms of difficult, or impossible to measure, properties such as internationalisation of rules, moral codes, social norms and reciprocity. He notes that determining an exact amount of trust would be very costly (if possible at all) and expects that trust will be measured as skill is measured: by indicators. Therefore Zucker (1986) proposes that trust could be explained only in terms of unmeasured antecedents: if rules are internalised – or moral codes or norms of reciprocity apply – then trust exists.

Actually, it is not quite clear if human beings consciously *measure* trust in their relationships. Parkhe (1993) borrows from Von Neumann and Morgenstern (1947), who have noted that people do not measure trust exactly, but rather conduct their activities in 'a sphere of considerable haziness'. Nevertheless, the results of this blurry measurement process are used in the everyday lives of human beings. If we agree with the premise of trust being a considerable economical and social lubricant (Arrow, 1974), *the ability to identify and measure trust* should be of great value both for scientific progress and practical 'know-how'.

Among others Parkhe (1993) and Koenig (1995) draw attention to the scarce literature on trust creation in contrast to the generous research on the collapse of cooperation and trust. Most economists do not believe in trust producing, but take it as either present or absent (Sabel, 1990). Zucker (1993) has, however, demonstrated 3 central modes of trust building: institutional-based, characteristic-based and process-based modes. His typology is used as a preliminary framework when searching for trust indicators in this specific context.

This paper is a preliminary attempt to find out how companies view the role of trust in a partnership formation context, and how they possibly attempt to create it. Research problems in this context could be stated as:

1. *How can trust be conceptualised in the context of small and large firm technology partnership formation?*
2. *How can trust be operationalised (in this context)?*
3. *What is the nature and role of trust (in this context)?*
4. *How do small technology firms attempt to create trust in their potential large partners?*

In accordance with these research questions, trust is first conceptualised. Thereafter an attempt is made to find suitable indicators for trust in this context. *The specific purpose of such exploratory work is to identify the means small technology firms use when attempting to create trust in partnership formation.* The perspective is a small firm perspective in the sense that it is not only the small firm's attempt to create trust that is focused upon, but also large firms are interviewed in order to evaluate their view of small firms' attempts in this area. However, a truly dyadic perspective on both parties' trust creation and assessment of a relationship and its formation is outside the scope of this paper.

Empirical evidence is provided by questionnaires and interviews. Two technology partnerships, where small high technology based firms have undergone partnerships with large, globally operating firms, have been examined by interview. Interviews have been open-ended, with the aim of finding some empirical perspective to the partnering phenomenon and the critical factors in it. Informants have been identified to be those intimately involved with partnership ventures from the beginning. In practice informants were small firms' managing directors and large firms' managers in charge of partner programs. Interviews followed an interview outline, but great freedom was allowed. The role of empirical material has been to gain understanding, set the phenomenon in its context, and find empirically based indicators for trust creation.

Additional material has been provided by a set of questionnaires collected during a high technology seminar. Ten persons, responsible for developing small technology firm sales, were interviewed about the means used to create an image of credibility and trustworthiness for their companies. In another seminar on technology partnerships between small and large technology firms, the participating 39 small high tech firms' representatives answered another questionnaire, where (among other things) they were questioned on their attitudes on partnering, and the role of trust in this process. All interviewed or surveyed companies have been operating in the field of information technology, and the small firms were specifically from the software sector.

CONCEPTUAL ISSUES

The term *partnership* refers here to the phenomenon in which 2 or more firms initiate a contractual non-equity cooperative relationship in order to utilise each others' complementarities. Co-operation refers to the definition by Buckley and Casson (1988) being 'coordination through mutual forbearance'. The contractual and non-equity partnership means that the parties involved have either a written or an oral agreement and do not own each other's shares. Partnership does not, however, involve a separate third party entity, most often referred as a joint venture.

The study is not limited to cases where both firms are active in technology development or marketing, but includes instances where one of the participating companies is active in marketing and another one is in technology. Also, in this work the technology partnerships addressed are *dyads* i.e. partnerships between only 2 firms. It is possible and probable that they operate in larger networks, but they are not a focus here.

Here the terms 'technology based' and 'high tech' are used to mean the same thing, which is that the company is based on its know-how and skills in emerging, new technology activities which are not yet widely available, and thus gives the owner (or owners) competitiveness and/or tradable assets (for a profound discussion on what is high tech, see Räsänen, 1994). For the sake of simplicity the term 'technology firm' is used instead of a high technology firm or technology based firm. In the empirical part of this paper, the companies studied are in the fields of information technology and computer software, which are generally accepted as high technology businesses.

The term *successful* partnership formation is used to describe a relationship in which negotiating parties reach a formal or informal agreement to form a partnership. Thus it does not refer to the consequent relative success of the existing partnership in monetary or other outcome terms.

According to the OECD standard, *a small firm* is defined as employing less than 100 employees and *a large firm* one employing more than 500. This definition by employment is approximate and incomplete, but hopefully gives the reader a clue to the differences between partners. In reality the studied small technology based firms were often much smaller, most of them employing less than 30 employees.

THE ARENA FOR SMALL AND LARGE TECHNOLOGY FIRM CO-OPERATIONS

Small technology based firms are of public interest and regarded as the innovating power within society. However, managerial limitations, limited access to skilled personnel, lack of finances and limited opportunities to achieve economies of scale complicate small firms' growth and survival potential. Small technology intensive firms seem to lack marketing skills, and attention, partly due to their scarce resources, is often concentrated upon R&D. Co-operation with a stronger counterpart, which has complementary resources, may help them to overcome some of the problems of smallness and youth. In addition to access to markets and financial resources, small firms are looking for larger firm collaborators in order to tap the reputation, status and the legitimacy of the bigger established firm. Potential advantages of a distribution-type co-operation include rapid access to existing markets, links with key buyers, and knowledge of local markets and cultures. Through co-operation with an OEM (original equipment manufacturer), small firms may also enjoy the benefits of a recognisable brand name (Contractor and Lorange, 1988; Oliver, 1990).

The behaviour of small technology firms does not seem to fit with wide-spread theories of the multinational enterprise (e.g. stage development or evolutionary models). Oviatt and McDougall (1994) aim towards a theory of international new ventures, which they define as 'business organisations that, from inception, seek to derive significant competitive advantage from the use of resources and the sale of outputs in multiple countries' (p. 49). Many technology based small firms are, in particular, 'international from inception' due to the homogenised international markets, fast and efficient low-cost communication technologies, and sometimes, internationally experienced and alert entrepreneurs. A major feature that distinguishes new ventures from established organisations is the minimal use of internal control and the greater use of alternative transaction governance structures. Due to their poverty of resources and power, new ventures may even use such structures when the risk of asset expropriation by the necessary external partners is high (Oviatt and McDougall, 1994).

Traditionally, many *large firms* have seen co-operative arrangements as a second best alternative to the strategic option of going-it-alone. Recently, a fast technological development pace, shortened product life cycles, versatile process technologies, falling profit margins, and the increased costs of R&D, have made the co-operation alternative more attractive. At present, many large firms are looking for 'windows' on emerging technologies, for example, in the rapidly developing fields of biotechnology and information technology the formation of partnerships between large and small firms are common. Of the 4200 European R&D partnerships identified by Maastricht Economic Research Institute on Innovation and Technology, information technology accounted for 41% and biotechnology for 41% (Shipman, 1992).

With regard to R&D, large firms may encounter problems of bureaucracy, internal inertia and risk aversion, but also often lack the dynamism and flexibility necessary for the initiation and successful conclusion of radical innovations (Segers, 1992). For a large manufacturer and distributor, a partnership with a small firm can be a less risky, more flexible, and a faster way to gain access to innovative new technologies (Forrest, 1990; Quinn, 1992). Actually, even multiple partnerships for risk-diversification purposes are found in many high-tech areas where companies may have a stake in many ventures with several potential competitors, in several technologies at various stages of development. Miles and Snow (1986, in Segers, 1992) liken this phenomena to dancing with multiple partners, or maintaining a 'loose network' where the strategy is to retain a stake (or potential payoff) from several, sometimes speculative, projects. According to Kogut (1988) the ability to carry out complex tasks, such as high technology R&D projects, is embedded within the organisation structure and within individuals in each organisation. This ability cannot be exchanged through markets, or even licensing agreements. In such cases, mergers or cooperative relationships may provide a more efficient form for complex inter-organisational learning (Moller and Wilson, 1988).

THE NATURE OF TRUST

Trust is usually based on an individual's expectation of how another person will perform on some future occasion; as a function of that person's current and/or previous behaviour. Trust is always perceived by the beholder, who makes a subjective assessment of the other party. Uncertainty, vulnerability, and the tendency towards risk-avoidance, as well as the need for choice based on judgment, are seen as obstacles which must be overcome if trust is to exist. In order to trust a potential partner, information is required. In conditions of perfect information, there would not be a question of trust, but rational calculation would be possible. However, if there is no information, faith or gambling is necessary.

Trust is based on experiences and social learning, and is thus also somewhat attributable to a specific local or national culture. It is more the property of collective units, rather than of isolated individuals. Therefore it is suggested that it is not the individual, but the relationship that should be the unit of analysis in trust research. Studying only individual persons (or firm) has been identified as a major drawback of contemporary research on trust (Lewis and Weigert, 1985; Salmond, 1994).

There is a strong temporal dimension to trusting (Halinen, 1994). Trust between partners is said to be a bridge between their past experiences and anticipated future behaviour (Salmond, 1994). Usually trust is seen as an outcome of a process, as trust relationships develop gradually. However, reputation and first impression are very important as, in many cases, one never gets another chance to prove his or her trustworthiness (Monsted, 1994). Probably the relative importance of past and future on trust alters over time, since in the relationship formation phase there might not be much past experience. As the relationship evolves however, the parties gain insight through experience and are thus able to form a better informed estimation of the trustworthiness of the other party. Thus trust can, at once, be seen as both a dynamic and static process. The level of trust in a relationship is constantly undergoing change, since it might grow or wither and constantly need to be rebuilt (Halinen, 1994). The culture of organisations evolve and change, and their constituent humans may change their behaviour due to negative experiences and failed estimation on the other party's trustworthiness. The process of trust building is seen as a self-enforcing process, in that trust creates trust, while distrust creates distrust.

Trust is most fragile and difficult to initiate, but is also slow to grow while always easy to break. Once broken trust is difficult to heal. Trust might appear either in brief encounters, as for example between strangers, or in long-term relationships. It may be both personal or impersonal, and indeed institutional levels of trust have been identified. The question of whom to trust is pervasive in many volatile businesses, as is the capability to be trusted. Trust has been identified to be critical for partnership formation and the future success of the cooperative venture (Halinen, 1994). Various streams of the relationship marketing approach acknowledge trust as leading to constructive and cooperative behavior vital for long-term relationships (Young and Wilkinson, 1989; Morgan and Hunt, 1994). A technology partnership between small and large firms involves many uncertainties. Rapid changes and high risks concerning technological success, commercialisation and economic rewards are typical for technology based industries. Exchanges between technology based, inventive small firm and large distributor are characterised by uncertainty and complexity. Perceived or believed dissimilarities in values, goals, time-horizons, decision-making processes, cultures and the logic of strategies imply barriers to the evolution of co-operation (Doz, 1988). In some situations partners make relationship-specific investments (e.g. intangible and tangible human and physical assets specific to a particular partner) and become, to some extent, locked into such joint ventures. In such cases, transaction costs become particularly high. Moreover, companies engaged in close cooperative relationships exchange and share valuable information, which may not be safeguarded by secrecy agreements. The fear of becoming vulnerable to opportunistic acts which cause the loss of this information may actually block negotiations aimed at co-operation. Under these conditions the ability to trust the potential partner becomes a critical issue.

Trust is always situation-specific, which means that the context matters. No universal definition seems therefore possible. In this paper trust is defined as *'the actor's expectation of the other party's competence and goodwill'*. This definition includes both competence (i.e. technical capabilities, skills and know-how) and the more abstract notion of goodwill, which implies moral responsibility and positive intentions towards a collaborator. It seems that very few researchers have distinguished the 2 different bases for trust, i.e. the here noted competence and goodwill dimension (for further discussion of the concept of trust see Blomqvist, 1995).

OPERATIONALISING TRUST

Operationalisation is usually understood as finding the operational definitions of theoretical constructs. In this technology partnership context the concept of trust was defined as a two-dimensional concept which was 'the actor's expectation of the other party's competence and goodwill'. *Competence* could be further operationalised via a set of questions referring to the conceptual parts of competence including technical capabilities, skills and know-how. The same process could be achieved in terms of the more abstract concept of *goodwill*, which again involves the conceptual sub-components of

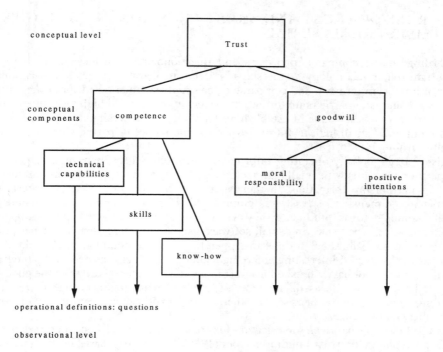

Figure 8.1
A preliminary framework of the transition of the concept of trust from a conceptual level to operational level

moral responsibility and positive intentions towards the other (see Figure 8.1).

This conceptualisation is made on a theoretical basis, in accordance with an interdisciplinary literature review, analysis and logical deduction. This framework serves as a good point of departure, but it must be made subject to empirically-based, and a more profound, investigation.

This is achieved by first reviewing how trust and trust creation have been operationalised by previous empirical research (see Appendix 1). The reviewed studies are mainly set in the buyer-seller and distribution channel context and all but one is of a quantitative nature. However, it should be noted, that very few of these studies gave any explicit definition of trust, and it was not very clear how the operational indicators were built (i.e. the implicitness of operationalisation might be a more common feature in empirical studies). However, this theoretical material will be used as a point of departure when searching for operational definitions of trust, and indicators for trust creation.

Secondly interviews and questionnaires on trust creation have been used to find operational measures for trust creation. This empirical material is seen as an important way of reaching validity. These indicators come directly from practice and put the issue of trust in its operational context. Both a qualitative research approach and technology partnership context make this attempt somewhat deviant from previously reviewed literature. As the concept remains vague and is little studied, the qualitative approach seems most suitable at this early stage. Qualitative method emphasises processes and meanings that are not rigorously examined or measured (see Denzin and Lincoln, 1994).

THE ROLE OF TRUST AND TRUST CREATION IN THE LIGHT OF EMPIRICAL MATERIAL

According to the company interviews and questionnaires, trust seems to have an important role in technology partnerships between small and large firms. Firms ranked trust, mutual commitment, and comparable goals highest when asked about the critical factors for succesful partnership formation. Seven out of 39 small high technology firms found trust a critical factor in establishing technology partnerships with large firms. Also 9 out of 37 small high tech firms explained the failure of partnership negotiation resulted from a lack of trust[1].

Also 2 large globally operating information technology firm managers, responsible for partner programs with small high technology firms, both stressed the importance of trust in partnership formation. In their small software partner selection criteria the competence dimension of trust was emphasised. First and foremost they evaluated small technology firms' product, if they existed, and attempted to evaluate their technical competence, when selecting small software partners (on large firms' partner selection criteria, see Tables 8.3 and 8.4 in Appendix 2). If the potential small technology firm partner possessed interesting emerging technology, which was needed by the large partner (but did not have a finished product or significant references), then the goodwill part of trust was emphasised. The role of goodwill trust was also highlighted as a deciding factor in those cases where there were small partner candidates of equal technology competence.

In the following passages some factors for trust creation are listed in accordance to Zucker's typology on trust building factors (1986). Zucker has classified the means of trust building as institutional-based factors, characteristic-based factors and process-based factors. Here the means for small firm trust building in technology partnership formation context are collected mainly as a result of 'open-ended' theme interviews with 2 small technology firm managing directors, and their above mentioned large counterparts' managers responsible for the relevant partnering programs. The first set of interviews were performed in 1993 and these were complemented by a second pair of interviews in 1996. Interviewed managers had been informed that the researcher was interested in studying how internationally oriented small high tech firms have succeeded in partnering large firms[2]. Thus, in interviews, trust as a theme has not been directly raised, but the interviewer 'fished around' the issue in order to explore trust, and what kind of role it had played in the partnering process. In addition to these interviews, 10 small high technology firm managing directors answered a questionnaire on trust creation[3], and these answers have also been included. A synthesis of the interviews and questionnaires mentioned above can be seen in Table 8.1.

Zucker's typology seemed to function quite well as an interview guide and platform for classification. Thereafter these factors are re-organised and classified in accordance with the defined conceptualisation of trust. The preliminary framework (see Figure 8.1) for operationalising trust did not function well as such, but a modified framework can be seen in the next Figure 8.2.

All the collected items for trust creation are included as 'operational' items for trust building. The conceptual components of competence and goodwill seem to remain valid, but the conceptual parts of competence have been modified to more descriptive classes of *inference on legitimacy* (close to Zucker's institutional based factors), *technical capabilities and know-how* and *business skills and know-how*. The conceptual parts for goodwill (i.e. moral responsibility and positive intentions) are left unchanged, as in the preliminary model.

Table 8.1: *A summary table on factors associated with small technology firm trustworthiness when establishing partnerships with large internationally operating distributors (in accordance with Zucker 1986 typology)*

I. INSTITUTIONAL-BASED FACTORS

- close contacts to acknowledged universities and R&D institutions
- key person's own scientific research
- company board
- venture capital or other well-known investors
- well-known partners
- memberships
- quality certificates
- location
- level of education

II. CHARACTERISTIC-BASED FACTORS

a) at company level
- references (known customers, difficult solved cases etc.)
- background facts on stability of the company (turnover, profitability)
- commitment (the product area or R&D)
- reputation for cooperating and being a trustworthy partner
- international orientation of the company (e.g. export sales)
- company visits
- experience of international co-operation
- high quality sales support material

b) at personal level
- professionalism in technology and relevant field (references of both key person and staff)
- personal credibility (e.g. expertness, positive intentions, personal attraction and dynamism)
- business-wise competence and salesmanship
- communication & presentation skills
- foreign and business language skills
- knowledge of international behavioural norms

III. PROCESS-BASED FACTORS

- first impression
- previous exchanges
- repeated encounters & personal contacts
- open communication
- initiative taking (e.g. helping partner, giving something extra)
- similarity of values
- similarity of goals
- proactive behaviour and support (e.g. hot line, quick response to questions)
- keeping promises

CONCLUSION

This has been an exploratory pilot study and a more profound study is needed. The trust creation factors identified seem however to have practical value in 'partner marketing' when relatively unknown, young technology firms, with little track record, attempt to persuade large potential firms to co-operate with them. It is assumed that small high technology firm managers could use these means more explicitly in their

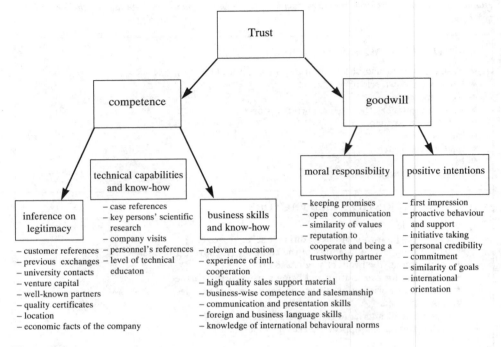

Figure 8.2
A modified framework for the transition of the concept of trust from a conceptual level to operational level in the context of small technology firm partnering

marketing and partnering efforts. Probably the means which could be used to signal trustworthiness towards large firm partners and new customers are very much alike.

In the light of the empirical context, the conceptualisation of trust as 'the actor's expectation of the other party's competence and goodwill' appears logical. It seems to include the means for trust creation and is yet parsimonious and clear. The original sub-components of the competence component were redefined in accordance with empirical investigation to small firms' practices on trust creation. This 'bottom-up' method in testing the construct validity seemed to work fairly well and has supported the conceptualisation.

A further research question enquired, 'what is the role of trust in this context?' However, this study is only able to give a tentative and partial answer which is that trust creation is used as important part of small technology firms' marketing efforts and large firms also note its value. The interviewed small firm managing directors highlighted how they attempted to create a trustworthy image by various means. The large firms' directors emphasised the more technology oriented competence indicators of trust. This pilot study does not, however, say anything about the role of large firms' trustworthiness, nor the means they possibly use when attempting to create trust. The dyadic process of trust building is also of interest, and could be studied as an idiographic case.

Finally, this study was able to show some tentative stimuli of trust creation in accordance with a conceptual definition for trust. These preliminary results encourage the further study of these issues in a way that builds upon the above tentative results and ideas, but seeks to overcome some possible drawbacks of the current research design. It is hoped that this modest effort is a beginning of theory development on trust and trust creation.

NOTES

1. This questionnaire consisted of 11 questions and was presented in 3 seminars organised by large IT firm A as a part of their campaign to find new software application partners. The title of the questionnaire was *'Internationalising Small Finnish High Technology Firms' Needs and Experience on Cooperating with Large Firms'*. All participating firms received the questionnaires and 39 returned those to the researcher.

   In the 1st open question, *'Which factors do you find critical in establishing technology partnership between small and large firms?'*, 39 small high technology firms answered this question by giving 47 different factors, which were classified to 10 classes. Compatibility of the firms got 16 remarks, small firms' product and competence 8 remarks and trust 7 remarks. Trust was remarked as 'Trust and belief', 'Mutual trust and compatible products and know-how' and 'Mutual need and trust'.

   In the 2nd open question, *'What if the attempts to establish a partnership do not succeed, what do you think are the reasons?'*, trust was named by 9 firms, lack of real need, different culture by 8 firms and product and small firms' lack of resources by 7 firms out of 37 firms that answered this question.

   In the 7th question, where firms were asked to *rank various factors* (familiar persons, compatible resources, good information flow, trust, compatible strategies, mutual dependency, similar company cultures, suitably tight agreement, suitably loose agreement, good personal chemistry, mutual commitment, mutual learning, compatible goals, similar size, existing business plan for the venture, openness and agreement on schedules) *in accordance with their importance to successful partnership formation between small and large firms* with the scale of 1–5 trust, commitment and compatible products were all ranked as most important and received an average of 4.5.

2. In addition to the basic questions on reasons for partnering, complementarities etc. questions to the small firm have been of the following nature: Why did the large partner choose you? How did you present yourself and your company to the other party? What kind of information did you give? Did you highlight some specific aspects of your firm? How should a small Finnish technology company market themselves to a large partner?

   Questions to the large firm have touched upon similar themes: e.g. Why was this specific firm chosen? How did this company appear? Why did they attract you? What factors do you emphasise while choosing small technology partners? What criteria do large firms in general use when selecting interesting but young and inexperienced high tech partners?

3. Ten small technology firms' managing directors and other key persons who participated a 'High Technology marketing' seminar in 1995 listened to the author's lecture on Technology Partnerships and the role of trust. After the lecture they complemented 2 sentences; 1) 'I try to create a competent image of company by following means . . .' and 2) 'I try to create a trustworthy image of my company by following means . . .'

REFERENCES

Arrow, K.J. (1974) *The Limits of Organisation*, WW Norton & Co, New York.
Blomqvist, K. (1995) The Concept of Trust, an Interdisciplinary Literature Review and Analysis, Research Report No. 2, Department of Business Administration, Lappeenranta University of Technology.
Buckley, P. J. and Casson, M. (1988) A Theory of Co-operation in International Business, in Contractor, F.J. and Lorange, P. (eds.) *Cooperative Strategies in International Business*, pp. 31–55, Lexington Books, MA.
Contractor, F. J. and Lorange, P. (1988) Why Should Firms Cooperate? The Strategy and Economics Basis for Cooperative Ventures, in Contractor, F.J. and Lorange, P. (eds.) *Cooperative Strategies in International Business*, pp. 3–31, Lexington Books, MA.
Denzin, N.K. and Lincoln, Y.S. (1994) *Handbook of Qualitative Research*, Sage Publications, CA.
Doz, Y. L. (1988) Technology Partnerships between Larger and Smaller Firms: Some Critical Issues in Cooperative Strategies, in Contractor, F. J. and Lorange, P. (eds.) *International Business*, pp. 317–38, Lexington Books, Mass.

Forrest, J. E. (1990) Strategic Alliances and the Small Technology-Based Firm, *Journal of Small Business Management*, 28, 3, July.

Forrest, J. E. and Martin, M.J.C. (1992) Strategic Alliances between Large and Small Research Intensive Organisations: Experiences in the Biotechnology Industry, *R&D Management*, 22, 1, pp. 41–53.

Håkanson, L. (1993) Managing Cooperative Research and Development: Partner Selection and Contract Design, *R&D Management*, 23, 4, pp. 273–85.

Halinen, A. (1994) Exchange Relationships in Professional Services, A Study of Relationship Development in the Advertising Sector, dissertation, Publications of the Turku School of Economics and Business Administration, Series A-6.

Koenig, C. (1995) Emerging Co-operation: The Dilemma of Trust vs. Calculativeness, a paper presented at the EGOS Colloqium in Istanbul, July 6th–8th.

Kogut, B. (1988) A. Study of the Life Cycle of Joint Ventures, in Contractor, F.J. and Lorange, P. (eds.) *Cooperative Strategies in International Business*, pp. 169–87, Lexington Books, MA.

Lewis, D. J. and Weigert, A. (1985) Trust as a Social Reality, *Social Forces*, 63, 4, June.

Meldrum, M.J. (1995) Marketing High-Tech Products: the Emerging Themes, *European Journal of Marketing*, 29, 10, pp. 45–58.

Möller, K. and Wilson, D.T. (1988) Interaction Perspective in Business Marketing, An Exploratory Contingency Framework, Institute for the Study of Business Markets, Report 11–1988, The Pennsylvania State University.

Monsted, M. (1994) Paradoxes of Complementarity and Trust, Problems of Network Export Strategy for Small Firms, a paper presented in The International Council for Small Business Conference.

Morgan, R.M. and Hunt, S.D. (1994) Commitment-Trust Theory of Relationship Marketing, *Journal of Marketing*, 58, July, pp. 20–38.

Oviatt, B.M. and McDougall, P.P. (1994) Toward a Theory of International New Ventures, *Journal of International Business Studies*, First Quarter, pp. 45–64.

Oliver, C. (1990) Determinants of Interorganisational Relationships: Integration and Future Directions, *Academy of Management Review*, 15, 2, pp. 241–65.

Parkhe, A. (1993) *Trust in International Joint Ventures*, a paper presented at the Academy of International Business Meeting, Hawaii, October.

Quinn, J.B. (1992) *Intelligent Enterprise: a Knowledge and Service Based Paradigm for Industry*, Free Press, New York.

Räsänen, H. (1994) High-Tech Knowledge – Buying the Invisible: An Enquiry into Buying International, Industrial, Complex, Knowledge-Intensive, High Technology Systems Products, Publications of Turku School of Economics and Business Administration, Series D-3.

Sabel, C.F. (1990) Studied Trust: Building New Forms of Co-operation in a Volatile Economy in Pyke, F. and Sengenberger, W. (eds.) *Industrial Districts and Local Economic Regeneration*.

Salmond, D. (1994) Refining the Concept of Trust in Business-to-Business Relationship Theory, Research & Management, a paper presented at the AMA Conference.

Segers, Jean-Pierre (1992) Strategic Partnering Between New Technology Established Firms and Large Established Firms in Biotechnology and Micro-electronics Industries in Belgium, MERIT 92–010, Maastricht Economic Research Institute on Innovation and Technology.

Shipman, A. (1992) Technology Inventors Make Out, *International Management*, September, pp. 50–3.

Young, L.C. and Wilkinson, I.F. (1989) The Role of Trust and Co-Operation in Marketing Channels: A Preliminary Study, *European Journal of Marketing*, 23, 2, pp. 109–22.

Young, L., Glaser, S. and Wilkinson, I. (1992) An Exploration of the Dimension of Interfirm Trust and Co-operation, a paper presented at the 8th IMP conference.

Zucker, L.G. (1986) Production of Trust: Institutional Sources of Economic Structure 1840–1920, *Research in Organisational Behavior*, 8, pp. 53–111.

APPENDIX 1

Review of the trust indicators and items for trust building in previous research

Author	Definition	Trust Indicators	Trust Building Items	Context
Bialaszewski and Giallourakis 1985	No explicit definition. An attitude, a personality characteristic.	• communication	giving enough accurate and useful information	An empirical study on channel managers' communication skills and resultant distributors' trust perceptions, 37 distributors were surveyed.
Lewis and Weigert 1985 (borrow Parsons 1963)	No explicit definition. Cognitive, emotional and behavioural dimensions of trust are identified.	• action for common values • common goals • expectations within solidarity involvement • relevant empirical information		Context of professional-client relationships, where competence or integrity of the professional cannot be validated.
Swan et al. 1985		• demonstration of dependability /reliability • honesty	setting expectations, use of proof sources and candid communication proofs, businesslike behaviour, bringing up also cons, and truthful admittance of limitations	A theory-in-use, 42 medical salespeople were surveyed on how they were able to gain customer trust. A general normative model of trust building as a result of the paper.
		• competence • motivation of customer orientation	technical knowledge, demonstration, proof sources and explicit statements assertion of customer benefits and unique requirements, stressing availability and bringing up product limitations	

		• likability	friendly behaviour, establishment of a common ground and businesslike behaviour	
Zucker 1986	Trust as a set of expectations shared by all those involved in an exchange.	• process-based • characteristic-based • institutional-based	past or expected exchange (reputation) personal characteristics (background) formal structures (certificate)	Historical study of the institutionalisation of trust in US industry in the period of 1840 to 1920.
Anderson and Narus 1990	'The firm's belief that another company will perform actions that will result in positive outcomes for the firm as well as not take unexpected actions that result in negative outcomes'.	• communication • co-operation • outcomes given comparison level	joint annual planning meetings performance objectives in order to establish expectations regular communication of performance specification of roles & efforts sharing market information retrospective performance evaluation	An empirical study of the distributor and manufacturer working partnerships. A dyadic perspective was used and trust was modelled as an outcome of co-operation.
Ring 1992	No explicit statement of the definition.	• trustworthy reputation • norms of equity • norm of reciprocity • experience on successful transactions in the past	No indicators given.	Theoretical discussion and presentation of proposals of the relationship of trust and contracts on organisational design.
Young et al. 1992	No explicit statement of the definition.	?	?	A general questionnaire to measure interfirm relations from both buyer and seller perspectives. 367 relations were studied.

Author	Definition	Dimensions	Factors	Description
Moorman et al. 1993		• researcher interpersonal characteristics	researcher's perceived integrity researcher's perceived willingness to reduce research uncertainty by interpretation researcher's confidentiality researcher's competence researcher's timeliness researcher's power	A theory of trust in market research relationships tested by 779 users. Theory focuses on factors that determine user's trust in their researchers.
Halinen 1994 definition from Anderson and Weitz 1989	Trust as one party's belief that its need will be fulfilled in the future by actions undertaken by the other party.	• specific trust: perceived confidence in parties' professional capabilities & perceived confidence in the other party's intentions regarding interaction reflected in opportunistic or reliable behaviour • general trust: estimation of the mutual confidence before parties came into contact	outcome satisfaction of previous assignment processes strong personal relationships interfirm knowledge common clarified interests common inter-firm roles cooperative interaction orientation indirect personal contacts company reputation	A longitudinal case study and theory based process model of the development of advertising agency-client relationships.
Haugland and Reve 1994 definition from Lewis and Weigert 1985	'trust as a social organisation' ... (whose) members act according to and are secure in the expected futures constituted by the presence of each other or their symbolic representations'	• norms • personal relationships • contractual solidarity	party's obligation not to harm the other party knowledge of the other party attempts to solve conflicts by working together	A theory-based and empirically developed model in distribution channels. 18 exporters answered a structured questionnaire.

Morgan and Hunt 1994	'one party has confidence in an exchange partner's reliability and integrity'	shared values communication	stress of corporate values communication of valuable information on expectations, marketing intelligence and performance avoid opportunism (superior offerings provided)	The authors develop a model that positions trust (and commitment) as key variables that mediate outcomes favourable to relationship marketing success. An empirical study of 204 independent tire retailers and their suppliers.
Wolff 1994	No definition offered.	• friendship encouragement • communication facilitation • gradual commitment	sharing social and business time sharing relevant information intercompany meetings frequent communication at various levels (e-mail etc.)	Practically oriented management interviews and normative advice.
Meldrum 1995	No definition offered.	• technology credibility • organisational credibility	lead user endorsement excellent media relationships high profile employees partnerships	A review of high tech marketing challenges and emerging themes for subsequent research.
Zaheer and Venkataram 1995 in defining trust they borrow Anderson and Narus 1990	Not explicit. Trust reflects 'the extent to which negotiations are fair and commitments are upheld'.	• mutuality • behavioural elements • process dimensions	existence of mutual trust fair behaviour reliable information	Theoretical model of relational governance is tested on data collected from a sample of 329 independent insurance agencies.

REFERENCES TO APPENDIX 1

Anderson, J.C. and Narus, J.A. (1990) A Model of Distributor Firm and Manufacturer Firm Working Partnerships, *Journal of Marketing*, 54, pp. 42–58.

Anderson, E. and Weitz, B. (1989) Determinants of Continuity in Conventional Industrial Channel Dyads, *Marketing Science*, 8, 4, pp. 310–23.

Bialaszewski, D. and Giallourakis, M. (1985) Perceived Communication Skills and Resultant Trust Perceptions Within the Channel of Distribution, *Journal of the Academy of Marketing Science*, 13, 2, Spring, pp. 206–17.

Halinen, A. (1994) Exchange Relationships in Professional Services, A Study of Relationship Development in the Advertising Sector, Dissertation, Publications of the Turku School of Economics and Business Administration, Series A-6.

Haugland, S.A. and Reve, T. (1994) Price, Authority and Trust in International Distribution Channel Relationships, *Scandinavian Journal of Management*, 10, 3, pp. 225–44.

Lewis, D.J. and Weigert, A. (1985) Trust as a Social Reality, *Social Forces*, 63, 4, June.

Meldrum, M.J. (1995) Marketing High-tech Products: The Emerging Themes, *European Journal of Marketing*, 29, 10, pp. 45–58.

Morgan, R.M. and Hunt, S. D. (1994) Commitment-Trust Theory of Relationship Marketing, *Journal of Marketing*, 58, July, pp. 20–38.

Moorman, C., Deshpandé, R. and Zaltman, G. (1993) Factors Affecting Trust in Market Research Relationships, *Journal of Marketing*, 57, pp. 81–101.

von Neumann, J. and Morgenstern, O. (1947) Theory of Games and Economic Behavior, Princeton University Press, Princeton.

Parsons, T. (1970) Research with Human Subjects and the Professional Complex in Freund, P. (ed.) Experimentation with Human Subjects, Brazilier, New York.

Ring, P.S. (1992) The Role of Trust in the Design and Management of Business Organisations, a paper presented at the Annual Meeting of the International Association of Business and Society, Leuven, Belgium, June 14–16.

Swan, J.E., Trawick, F.I. and Silva, D.W. (1985) How Industrial Salespeople Gain Customer Trust, *Industrial Marketing Management*, 14, pp. 203–11.

Wolff, M.F. (1994), Building Trust in Alliances, *Research and Technology Management*, May-June.

Young, L., Glaser, S. and Wilkinson, I. (1992) An Exploration of the Dimension of Interfirm Trust and Co-operation, a paper presented at the 8th IMP Conference.

Zaheer, A. and Venkataram, N. (1995) Relational Governance as an Interorganisational Strategy: An Empirical Test of the Role of Trust in Economic Exchange, *Strategic Management Journal*, 16, pp. 373–92.

Zucker, L.G. (1986) Production of Trust: Institutional Sources of Economic Structure 1840–1920, *Research in Organisational Behavior*, 8, pp. 53–111.

APPENDIX 2

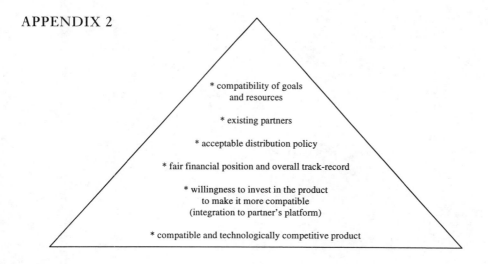

Figure 8.3
Large information technology firm A's selection criteria for their small software partners

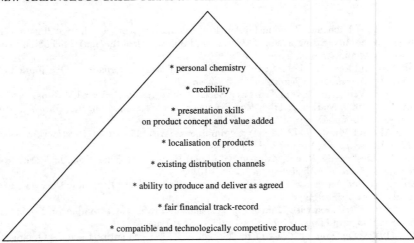

Figure 8.4
Large technology firm B's selection criteria hierarchy for their small software partners

Firms' selection criteria were somewhat similar, competitiveness of the existing product was the major criterion. Both interviewed managers responsible for partner programs emphasised the role of trust in their speech, but the competence-based trust criteria were clearly the most important when they evaluated potential partner's trustworthiness. Of the goodwill-based trust items commitment to technology was mentioned. In addition proactive behaviour and active support to sales and product technology were emphasised as a proxy of how the small firm would perform if delivering projects to large firms' customers.

CHAPTER 9

Strategic Alliances as an Analytical Perspective for Innovative SMEs

METTE MØNSTED

INTRODUCTION

This paper is based on an ongoing research project entitled High Technology Firms and Networks. The purpose of this work is to identify different forms of networks and forms of collaboration, with a focus on strategies and the process of networking.

Small new and innovative firms paradoxically appear to be constrained by the demand for documentation under conditions of innovation. The paper focuses on the conditions for innovation in small innovative firms. It is not intended here that this paper should provide a macro view of 'typical' general innovations, but address innovation problems of relevance to innovative small firm policy in particular.

The concept of strategic alliances is usually applied to large corporations, their concept of strategy, and their ability to manipulate their environment. This principle is based on a balance between firms in terms of complementarity. But it is also often used as a general vehicle for collaboration between small innovative firms as a means of creating 'economies of scale'. The concept of balance in this context is, however, different for the small firms, since they often have no physical assets, but only technical opportunities for innovation collaboration. The perspective of using networks strategically by small firms should both cover access to resources as well as obligations within alliances. However, this research is concerned, not only with networks, but also with the methodological problems of grasping the fluidity and ambiguity of these relationships, together with an investigation of the process of building different networks for innovation purposes.

The empirical basis for the paper is, at present, 13 case studies of small high technology firms. Two of these enterprises have been interviewed 6 times over a one year period in order to get more 'in-depth' knowledge of critical events within these firms, and to provide information on the changes that occur within networks. Interviews with 5 R&D engineers in the computing industry of Silicon Valley during 1993 created the basis for the formulation of this project, and for the emphasis on a more general perspective towards the contradictions and dilemmas felt by such development people when using personal networks. A constant problem for small firms is to establish their reputation, and prove their legitimacy. Improving their list of contracts seems to be closely dependent upon the strategic use of networks, which may involve some dependence on large customers, while guarding against the opportunistic behaviour of such large patrons, given their size and resources.

STRATEGIC ALLIANCES AS A CONCEPT

The use of a strategic alliances approach, however, reveals some contextual problems. In fact, this concept is a combination of 2 approaches, which derive from different paradigms and theoretical traditions. The 'strategy' component is closely tied to the classic business economic tradition (Annsoff, 1977; Hamel and Prahalad, 1994), and has been developed in the context of large corporations. It concerns choosing between many different options with regard to business products, markets and possible alliances. It is fundamental in most of the strategy literature, to maintain some kind of flexibility between options. However, the 'alliance' part of the strategic alliance concept originates from political science and, for example, might include the alliances that occur between countries at a certain stage of a defence pact. However, it is a concept now used in relation to industrial collaboration within industrial networks, and implies some kind of dependency or obligation between firms. Nonetheless, dependency can be either a positive or a negative issue in relation to the strategic alliance concept, since in addition to advantages, any alliance will contain mutual constrains and obligations.

The concept of strategic alliance has been applied in many contexts where it might be dismissed as just the current 'buzz word', or criticised as not contributing new insights during analysis. It has been used to describe many forms of collaboration, which might be networking, more formal alliances, or even joint ventures. In terms of definition, it is tempting to start from the negative viewpoint of indicating what is not a strategic alliance. For example sub-supplier relations, or informal networks, might not be viewed as strategic alliances, but should probably be seen as a sub-alliance of networks (another unclear concept), here defined as series of formal and informal relations held between individuals, which create personal networks that are ego-centred.

A major problem with the literature is that one set of explanations is based on an economic transaction cost perspective, predicated on the 'winner–loser' concept (or winning under competitive conditions) (Jarillo and Stevenson, 1991; Lorange and Roos, 1992). However, a second set of definitions depart from the network and alliance perspective, stressing the basic idea of the 'win-win' situation (Johannisson, 1986; Johanson and Matson, 1988; Pyke, 1988). Yet other researchers try to create hybrids of the above approaches (Powell, 1987; Fletcher, 1993; Powell et al., 1994). This ambiguity, and the ensuing confused picture, mirrors real problems experienced in many firms involved in actual strategic alliances. In order to attempt a definition of the concept of strategic alliances as a form of alliance and/or networking, which differs from other collaborative forms, the strategic alliance might be defined as:

> A formal collaboration between two or more firms to obtain a target of mutual interest, to all concerned.

Other authors might hesitate to attempt formalisation, since this is often not the most important feature of an alliance, and indeed, the context and content of the alliance may change *during* a collaboration (Dodgson, 1993). However, it is important to avoid very marginal or irrelevant relations, although this does create problems when considering collaborations between large and small firms, where an alliance is often more important for the small than it is for the larger firm.

Some authors would be eager to include joint ventures as a form of alliance by implying that a dependency between the firms could be administered through a joint unit, or as a totally new separate firm. This was demanded in an industrial policy programme entitled 'The Network programme' in force in Denmark between 1989–92 (Industri- og Handelsstyrelsen, 1990). Other Danish authors, working with small firms, would not include joint ventures as strategic alliances, but would define them as a separate form of collaboration (e.g. Gustafsson et al., 1991). Fletcher (1993) focuses on horizontal relations by trying to find relationship forms outside the vertical customer-sub-supplier relationship. She also stresses the gradual development of alliances, and the *ex-post* label of a strategic purpose, which might not have been originally intended.

Explanations developed for strategic alliances are often constructed from the perspective of large corporations as strategies, but do not reflect the dependencies involved, or the processes involved in implementing a collaboration (Lorange and Roos, 1992). Their perspective is dependent upon market and hierarchy perspectives, between low and high integration, and between low and high dependency (ibid). How these 3 dimensions are related to the analysis of strategic alliance is not very clear. However, they are dimensions which would be fruitful to apply in an effort to clarify the strategic alliance concept as a model, and as a strategy for different types of firms.

Using combinations of the variables (or dimensions) noted above to define and evaluate the concept of strategy could be fruitful, when this general perspective is specifically applied to small innovative firms. Relevant dimensions for defining the model in this context include:

hierarchy	- market
high dependency	- low dependency
vertical (value added)	- horizontal
formalised	- not formalised
strategy	- gradually developing (process)
planned	- ex-post rationalisation
complementarity	- similarity

A problem of classification is the serious functional overlaps that exist concerning networks. Consideration of one role of a person may open up other roles, thus changing the relationship, but not necessarily warranting a new dimension. The above dimensions may help us understand the kind of trust and dependency inherent in a relationship. Some dimensions tie content to the purpose of the interaction, such as strategy and complementarity, while others to the form and the kind of relationship. The 'glue', or the strength, of the relationship applies not only to 'tight' relations (Granovetter, 1973), based on similarity, but also to relations based on complementarity, in terms of a division of labour.

The cohesiveness of a network could be expressed in terms of the Töennies concepts of Gesellschaft vs. Gemeinschaft used to explain the solidarity and integration of different societies (Töennies, 1957; Asplund, 1991). The Gemeinschaft is reflecting the closeness and integration formed by similarity (ibid), whereas the Gesellschaft reveal the need for complexity and variety in a modern society.

With the economic-strategic model as a point of departure, strategy must involve finding partners with complementary competencies and resources with which to form a stronger combined unit, without the weaknesses that would have occurred had one firm tried to fulfil all the competencies needed alone. Thus, the basic idea is not to find a firm with similar competencies, but to discover one with different assets. It is a fundamental part of this process that the firm seeking an alliance partner has some strengths or opportunities to offer and is able to play the role of 'strong player' in a partnership. This is a strategic perspective also discussed by Burt (1992), who considers the strategic network overcoming 'structural holes'. This bridging process in order to access other resources is, however, particularly relevant for small firms. It is a means of accessing external complementary resources, with which to augment internal resources, and achieve combined resources. This will create active players, but not necessarily a 'strong player' envisaged in Burt's advocation of the concept (ibid).

The basic models concerning the strategic use of networks are elegant, but when applied to small firms, the 'strong player' is difficult to find. Strategic alliances within this size category of firm are also difficult to identify, and more relationships seem to be of a consultancy type, rather than of a strategic alliance type. But some strategic alliances may be found. The types of collaborations found among small firms are more often gradually developing into significant relationships which, *ex-post*, could be defined as strategic alliances rather than the process prescribed by the theoretical model, where the strategy is first. The perspective of 'know-who' before 'know-how' seems to be important, in that access to new information is through people. This perspective on strategy is developed by Fletcher (1993) in relation to the concept of strategic alliances, where the definition of strategy is expanded to include alliances that gradually develop, where strategy is not a joint perspective developed from earlier network relationships.

STRATEGIC ALLIANCES AS A PROCESS

Alliances, and the creating of confidence and trust are easier to establish between firms and between persons with similar characteristics. In the initial stages of a relationship both complementary and difference factors are strengths, but as the process of collaborating and implementing the alliance proceeds, such differences may become a weakness. In these later stages, similarity and closeness are usually beneficial. Indeed, trust is a fundamental 'glue' of networks and informal alliances. Strategic alliances, however, may be formalised, and the relationship 'spelled out' in a contract to eliminate possible misunderstandings; but in essence, these alliances are formed in order to share information, to create openness, in relation to specific projects (Powell and Smith-Doerr, 1994, p. 388). It is a paradox that the best collaborations often occur in firms that are alike in all other aspects, apart from limited areas of difference, which are the basis for collaboration. This is one reason for observing that firms should be of similar size, know each other before the collaboration, be of similar technical culture, within the same age-group, or possess other features of similarity, as a basis for collaboration.

When evaluated in a process context, the network is based on ego-centred relationships, where the focus is on one firm, and the options of this firm are expressed in a network perspective. The focus includes an evaluation of whether there are barriers, and the identification of other relevant networks or groups, in which the firm does not have contacts. The interest is then concerned to finding out how to create access to other groups, and in particular, who should be contacted, and who might know 'these other people'. Burt (1992) has worked on this perspective as a means of determining how access is gained to a network, and how confidence is created in relevant network 'circles'.

Applying Burt's model (see Figure 9.1) in its crudest form to a small business context, however, may be hazardous, since small firms are very easy to 'bypass', since they are often marginal to many networks. However, as a model used to inform us about networks in terms of where small firms do not have access, it may be relevant to find out who should be contacted in order to get some kind of access to other new important networks.

A key issue here is that of autonomy and of avoiding constraints, which might only be the point of view of the very large and dominant partner of an unequal relationship. However, the weaker partner might try to search for some kind of obligation or constraint on the part of this stronger partner.

This rationalisation of network contacts could be also applied in terms of a 'mentor'-model in terms of finding a relevant person who could sponsor a small marginal firm with a view to creating relevant contacts. The model therefore changes perspective to

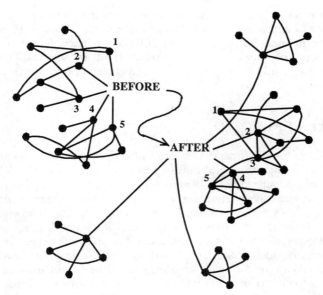

Figure 9.1
Strategic changes of networks to overcome structural holes
(Source: Burt, 1991, p.22)

one of strategic awareness about important potential contacts, which cannot be reached. But the paradoxes are obvious. By using a more powerful partner within a strategic alliance, as a mentor, this means of accessing networks is necessarily asymmetric and the power relationship is thus unequal.

Burt's model is partly intended to be an access and control mechanism in which intermediaries are used to control others. This concept would definitely be relevant in controlling a very powerful player, but possibly less so in other instances. The control function of a mentor, however, could be transformed to some other kind of influence, such as creating trust and legitimacy in order to recommend a client to others in a network.

This leads us to seek clarification of 2 different aspects of the 'networking' process. Clearly, network access is important in order to overcome barriers in order to obtain new information, and meet new persons. The classic article of Granovetter (1973) on 'The strength of weak ties' considers this by dealing with access to information from sources outside a group with strong ties. Burt is also concerned with this problem in order to overcome 'structural holes', and emphasises that it is not so much the strength of network ties, as much as it is the lack of overlap or redundancy, which is the basis for getting access to, and control of, *new* information.

Another aspect of control may become important for the powerful firm, which may play the 'emperor' that can 'divide and rule'. This firm may control sub-suppliers and make sure that they do not communicate and make alliances between themselves. This issue is important for some of these 'rulers' such as, for example, Volvo or Novo Nordisk who, as large corporations, maintain many sub-suppliers, over whom they wish to maintain a central control. The power relations in these networks thus become important. This approach to networking, however, departs totally from the fundamental view of the network as being based on trust. Such an arrangement is steeped in competition and mistrust, and thus forms part of a different paradigm. This paradigm is one of a strategic alliance, which creates synergies between collaborating firms, but which will also have to deal with *some* of the problems of opportunistic behaviour involving mistrust and competition. The basic idea of this strategic alliance, however, is different from the model created by Burt. The redundancy perspective is not strong, and is not necessarily negative, as this creates trust and motivation, and a kind of dependency which could be used in a continued collaboration, possibly aimed at either existing or new firms.

EMPIRICAL REFLECTIONS

The empirical basis for this paper is a series of case studies on innovative firms within biotechnology, the pharmaceutical industry and computer software. All these cases are within technically advanced fields, where industrial secrecy and protection are important issues. Interviews have been undertaken with owner-managers, project leaders, and development staff. Two cases, followed over time, involve owner-managers and project leaders and both are derived from small firms of 7 and 4 employees.

Small, and especially new firms, founded in science parks, are very dependent on external resources, which create trust based networks. The general idea of the science park is that collaboration and trust building should be within homogeneous networks and based on 'equality' (Cook, 1977; Burt, 1982). Due to the smallness of such firms, it may be difficult to create a proper balance in collaborations with others. Power and influence seem more to be attached to personal credibility in personal networks than to other formal criteria. They may be individually valuable for people who develop networks, but for firms, they are hardly powerful. At best such firms may, for a long time, represent opportunities for significant innovation, but with so many obvious weaknesses, their commercial potential could easily evaporate. For example, some of these enterprises may be too 'early stage' to be of interest to venture capital firms. This creates the problem of, what are the opportunities inherent in the development projects of such firms? This seems to be an especially important factor in the many collaborations between large and small firms within the biotechnology sector (Dodgson, 1993; Powell and Doerr-Smith, 1994).

The following discussion presents an overview of some of the choices that must be made in terms of units of observation and units of analysis:

Unit of observation (i.e. types of firm):

1. innovative
2. small
3. new

Unit of analysis (network attributes):

4. process more than structure
5. access
6. control
7. strategy
8. innovation project

(1) Innovative Firms

Studying innovation is, in this context, tied to product developments with a certain level of invention, and thus involves uncertainty. Using the Schumpeterian definition of innovation, innovations must have a new element (Schumpeter, 1934), but it is not easy to identify the economic value of such advances when they have not yet been launched on the market. In this context the market value can be seen as unclear. Thus strict economic value on the open market is very difficult to judge. Moreover, this innovation concept, not only includes product inventions, but allows for other organisational and market related innovations. Thus, innovation, by definition, includes many forms of technical, market and organisational uncertainties as an embedded feature of the total innovation process. Moreover, small firms do not have the brand names and corporation name to lend credibility to their projects. Small firms also depend on only one or a few competent persons. They do not only need to legitimise their inventions or innovations, but also must establish the fact that they are serious and permanent organisations or firms. Entry barriers may be very high, and if new enterprises do not succeed in getting some kind of access to a market, they may remain inventors rather than innovators.

(2) Small Firms

Being a small firm implies uncertainty. To what extent is this a 'firm' or just a very competent person? Is this a firm that customers would trust, or is the whole project totally dependent on one person? On a formal level, this is a serious problem for relations with, for example, financial institutions. When small firms do not have power, they must possess certain powerful connections in their networks to support the first stages of their life. They are very vulnerable at this point, and often need help from others.

(3) New Firms

New firms have less stable structures on which to build. Routines are fewer and they have to start up new access routes for networking. Entry to support networks and trust based relations are also rare. Often the behaviour of such enterprises is hard to plan for, or to rationalise, and it is tough to generalise beyond the point of noting their newness.

(4) Process More Than Structure

The concept of the network is a social construction, not always recognised by the actors concerned. When network information is tied to innovation projects, stable structures are less important than dynamic interactions. Thus the focus must be on the dynamics of key stages in the innovation process (Curran et al., 1993; Mønsted, 1995). In terms of the measurement of dynamics, there is a difference between the scope of change, as represented by the measurement between 2 points, and the rate of change. Breakthroughs may reveal information, not only on the change process, but also on the feelings of individuals before the change occurred.

One of the problems of measuring innovation and related network strategies is that of dynamics and unstable structures. The structures, and well documented experiences, may form a basis for trust and credibility, and provide an access vehicle to new people. The network, as a structure, is therefore both a result of network processes, as the dependent variable to be explained, while at the same time it is a path for access and a possible means of control, and thus an independent explanatory variable (Mønsted, 1995). Network theory is not clear on whether the study of network structures is aimed at the network *per se*, or a means for achieving other purposes.

Structure and stability may set the framework for the communication, and ensure predictability. But it is characteristic of really innovative projects that the early stages have a particularly high level of uncertainty. Novel insights, and their reliability, are based on a combination of tests, documentation and information from trustworthy people in that the chaotic image of uncertainty is gradually clarified through the acquisition of bits of reliable information. Early decisions on whether to continue collaboration or funding takes place at a stage where uncertainty is embedded.

(5) Access

An empirical study should include consideration of access to information, access to customers, or access to collaboration partners. The access to new information and persons is basically tied to a strategic perspective of where and how to obtain contact with relevant persons and groups of persons.

(6) Control

The control of information, and of social exchange relations, is interesting in terms of the application of informal forms of social obligations as control, rather than relying on legal forms. Social control and trust is relevant for the creation of legitimacy, involving trustworthiness, which is tied to social control and the recommendations of other persons. The social control of persons and information creates a basis for establishing other forms of the valuation of the quality of projects and firms, where information is very scarce and uncertain.

(7) Strategy

The use of the network as a strategy to get access to new and reliable information is an important technique, when information streams are extensive, both in terms of finding information on trustworthiness, and on complementary partners for strategic alliances. However, the structuring of data or 'virtual data' may be difficult. During the process of identifying the contexts in which information and contacts are needed, it is important to open new relationships, rather than support existing contacts. Strategies to find where information resides, and what kind of strategic alliances should be formed in order to provide access to resources while providing security and protection, remain important issues for research.

(8) Innovation Projects

Innovative processes occur at intervals during the firm's life. Following innovation, both periods of rapid growth and barriers to growth are experienced. The innovative project itself has to be studied in order to understand the processes, together with the roles of internal and external ties. In biotechnology, the process appears to take the form of more frequent collaborations between firms (Powell and Brantley, 1992).

Two case studies have been constructed over a period of a year to follow changes in alliances and the changed meaning of network contacts, a frequent feature according to Dodgson (1993). Following the changes involved in a network resulted in many dead-ends. This also involved a study of how access is created, and how different forms of connections are applied to create a basis for legitimacy, and as access to resources. Some researchers do not believe that networks represent a 'safety net' (Curran et al., 1993), while others claim that networks provide information and support based on trust, but do not include resources (Stinchcombe, 1989). In this process the creation of networks as investments in a support structure (i.e. strategic personal contacts for later use) are also included, and seem to be evident within the group of case study interviews reported below.

STRATEGIC ALLIANCE AS A PERSPECTIVE ON THE PROCESS OF NETWORKING

In the following 2 case studies, conducted over an extended time period, the type of problems encountered cannot be solved by one firm alone, but have involved collaboration between case study firms and other firms or other experts, in a context where Denmark is felt to provide a very small constraining market. People with expertise are few, and they exist in differing research environments.

In particular, in one case, the firm's expertise was demanded for projects in both Germany and the UK. In such forms of collaboration, where firms work together with customers, it could be questioned whether this is marketing, or whether it is networking, or some other form of collaboration. The selling of know-how in different long term arrangements is not merely a sales relationship.

During the first contact with new customers involving leading edge innovation, it is often difficult to persuade others that the small firm in question may be able to solve problems that large American companies have not been able to resolve. The basic problem is that they are both serious technologist and charlatans promising solution to innovation problems. The number of small firm owners making unrealistic promises raise doubts among customers, and make it difficult for serious small innovative firms to retain credibility. Although the image of the alchemist was credible for some time, a lack of results changed credibility to incredulity. However, the hope of finding 'gold', kept many venture capitalists of the day interested for a long time. However, the need for trust without much collateral can also be difficult to accept, and the balance between trust and documentary evidence on performance is a problem for most innovative firms in their early stages.

In the early stages of communication between an innovative firm and its customers, a high level of uncertainty makes the distinction between 'facts' and 'fiction' very difficult. What is real, what arguments are based on tested facts, or which are 'daydreams', is difficult to distinguish. Moreover, the secrecy necessary at this early stage also makes it even more difficult to make credible scientific arguments generally available to legitimise concepts. This is one of the fundamental paradoxes related to all innovation decisions.

The interviews conducted report on a number of innovators, who believe themselves to be geniuses, while they view customers and collaborators are idiots, who do not understand their genius. This 'attitude problem' inhibits information exchange and serious dialogue. From this difficult 'point of departure', it is hard to 'break the ice' and create credibility during communications. It resembles the problems known to service firms, who cannot demonstrate or test their services in advance of purchase by customers (Normann, 1984). The danger is that some highly innovative firms may be classified as an 'alchemist', and rejected as bogus at a very early stage. For example, a small biochemical firm had a meeting with the research manager of a large electrical corporation, specialising in cables. The small firms (at this stage theoretical) invention could considerably increase the capacity for conductivity in cables. However, the research manager of the large potential customer firm stated that 'they were not interested in the improved conductivity of cables'. The small firm owner-managers were surprised to hear this. The research manager obviously did not believe the small firm's claim, and did not want to waste time on this proposed solution. The image of being an alchemist, or maker of 'magic potions' is a serious barrier to acceptance and possible collaboration. The problem is also that, in such innovative projects, selling is a complex process, and must act as a basis for collaboration to exploit the further practical use of the invention, prior to market testing.

In another instance, a firm was trading as a computing firm while they sought to develop a new chemical innovation. Initially they identified a sector where they could promote the application of the new chemical invention, and started to explore the customer base through computing customers, embassies, and a few research colleagues at universities. However, exploitation was very difficult and frustrating, since they began to identify more and more types of firms to be potentially relevant, but had no network contacts with which to fully access them. They did not initially manage to make the right contacts, but kept trying, and were forced to 'kiss frogs in order to find the prince' (Munksgaard, 1995). The contacts were not very systematic, nor very seriously related to the technology of their scientific invention. A few 'contacts' via internet groups gained access to interested scientists, but without much access to commercial customers.

At this stage the firm identified the need for a friendly 'big brother' who might show the way, and finance development expenses. This is where the strategic alliances perspective emerged, as a kind of marketing model for young early stage innovations. The finding of partners for co-operation may be seen in this light. This same firm, in a third phase of networking, began to use its professional contacts, which were comprised of chemists and scientists working on innovative projects in research institutions and science parks. However, through this route they only found interested parties with no industrial contacts. This time, however, they also accessed new networks of relevant science contacts in the UK and Germany with access to large corporations. The firm and the scientific community, through the network, are currently accessing EU funds to supplement their economic base to help fund a long and difficult start up phase. But this also provides the legitimacy for new commercial and practical collaboration projects. Also at this stage, the firm needed to disclose information in order to find possible alliance partners. This disclosure of information to others is a very frustrating, but necessary process.

The documentation to prove the credibility of a new technology is hard to provide at an early stage in the innovation process. For example firms in the field of theoretical chemistry, may not have a physical prototype. The above mentioned firm is at such a stage, where most US venture capitalists would also 'wait and see'. However the scientific articles, and the personal credibility of scientists seems to be the access road to new foreign commercial ties. The process is now partly working, although some stages appear totally blocked. Only by having another commercial activity had the firm concerned survived for 2 years, by January 1996.

Radical inventions, at least within some of these sectors, are hard to achieve without public funding. EU supported projects are clearly an option, since they allow an otherwise non-viable commercial firm to be fundamentally supported to perform R&D in a research field, although there should remain a bias towards kinds of scientific-commercial activities where, eventually, there is hope of commercialising new technologies.

When the innovation process is analysed from a strategic alliance perspective, the need for mentors becomes apparent, and other (often larger) firms with complementary strengths may provide help. However, in Denmark, where venture capital for innovative small firms is practically non-existent, capital market inputs should come from national government, EU funds, or from industry.

The model of strategic alliance is synonymous to a kind of 'balance of power' between partners. The related parts of this network would stress the creation of trust among equals (Blomqvist, 1994), but in innovative small firms, the focus is on large firm partners, who have more resources than the small firms. Access to resources in the evaluation of complementarity seems to be more important than the danger of inequality and/or dependency. Even if problems of dependency become an important issue, the need for resources tends to override these dangers.

At this stage, however, conservative and steady development may kill the innovation, and often the dilemma is that, either you collaborate with the large corporations, or your project will die. The strong need to find possible partners is evident, and many small high technology firms cannot afford to be very critical. Some of the firms maintain their autonomy much longer, since they have a kind of 'cash-cow' in the form of an economic activity which can earn their living while they are building up the innovative activity.

While most strategic alliance models are based on a balance between strong firms who can choose from various options, the model changes radically when applied to small innovative firms. Whereas the strategic element in the networks (strategic alliances) of less innovative small firms are defined *ex-post*, and develop through the networking process, many small innovative firms are *forced* to adopt strategic alliances, either as customers or as business partners in a range of different kinds of alliances. But when a large firm is the target, is this to be considered an alliance for the large firm concerned? Is such a relationship strong enough to be considered an alliance, or is it merely a normal customer-supplier linkage? Indeed, what do we do with strategic alliances where only one partner considers the alliance strategic?

CREDIBILITY AND THE EVALUATION OF DEPENDENCY

In the cases discussed above, a lack of credibility in the early stages of alliance contacts was a serious problem. Whereas large firms create trust and legitimacy via their size, brand name and/or documented innovative projects, small firms, and especially new small firms, do not have this option. Small firms have to prove themselves worthy and valuable. The creation of trust, and the persuasion of customers and partners that they are a legitimate innovative firm is a process involving, not only the firm owners, but wider personal contact networks.

The early stages of an alliance in which partners 'test' each other, is not as simple as a contract-based model. While formalisation may become important, the combination of personal relations, person-based recommendations, and bits and pieces of scientific information to examine the ability of the persons, initially comprises a loose system. In this system, scholars openly admit to uncertainty, the need for different methods of evaluation, and risktaking. Americans have argued that, if they followed the lawyer based contract model, this would block opportunities, although some of the contacts proved to be dead ends or 'alchemists'.

The strategic alliance model, as a static model, is hard to use to describe innovation and strategic perspectives for small firms. But it seems to be useful as an inspiration and 'role-model'. The model does not, however, reveal any tools, or help in relation to the many ambiguities and dilemmas created by the embedded paradoxes of the model. This makes it difficult and dangerous to use, since the paradox of, for example, leaking information vs. industrial protection is embedded in the whole process of innovation and the creation of alliances for small firms.

Dependency and constraints are negative concepts in terms of more economic strategic thinking. Dependency in network theory is interpreted from negative, redundancy (Williamsson, 1975; Burt, 1992) to mutual obligations (Cook, 1977; Johannisson, 1986; Aldrich, 1987). These differences are fundamental to the 2 paradigms associated with strategic alliances and networks, ensuring that the interpretation of linkage dependencies is extremely difficult.

A player in large firms wants to maintain all kinds of options and possible alternatives, and would like to keep information away from smaller firm partners in a 'divide and rule' manner, giving him power over dependent sub-suppliers or collaborators.

However, the small firm view is different, as this is not only tied to the problems of dependency and power relations, but also to the use of the network as a means of forming ties of obligation. Such obligations are governance structures intended to hinder large firms from involving competing collaborators. Whereas the obligations between equals may have this loyalty characteristic, it is harder to find between larger firms and their small firm suppliers, as this type of obligation is based upon individuals, who may move between positions in large organisations. However, the effort to create 'glue' in the network, by trying to create the personal relations of this 'obligation' type, is an important strategic option for small firms.

CONCLUSION

The concept of strategic alliances has been fashionable for a long time, and thus has entered the vocabulary of consultants as a general 'miracle-win-win' model tied to the successes of large American collaborating companies. However such behaviour is often close to being accused of creating monopolies; for example when Microsoft's Bill Gates creates alliances in the television and/or film industries. This is very different, seen from the position of weak small firms seeking strong partners. One of the problems is that the same vocabulary is used in both instances, although the roles, dependencies and power balances are totally different. The search for legitimacy via partners within networks is a part of this process. Legitimacy and proper references are more important than anything else under high levels of uncertainty during the early innovation phase of a small firm.

The analysis of cases over a long period of time confirms the methodological problems of analysing networks, and that networking process, rather than the structure of networks, is the important factor regarding the success of innovation projects.

REFERENCES

Aldrich, H., Rosen, B. and Woodward, W. (1987) *The Impact of Social Networks on Business Foundings and Profit: A Longitudinal Study*, Paper presented at the 1987 Babson Entrepreneurship conference, Pepperdine University.
Ansoff, I. (1977) *From Strategic Planning to Strategic Management*, Wiley, London.
Asplund, J. (1991) *Essä om Gemeinschaft och Gesellschaft*, Bokförlaget Korpen, Göteborg.
Blomqvist, K. (1994) *Trust in Technological Partnerships – Context and Conceptual Issues*, Nordic Workshop on Interorganisational Research, Ålborg, Aug 25th–27th.
Burt, R. (1982) *Toward a Structural Theory of Action*, Academic Press, New York.
Burt, R.S. (1992) *Structural Holes: The Social Structure of Competition*, Harvard University Press, Cambridge, MA.
Cook, K.S. (1977) Exchange and Power in Network of Interorganisational Relations, *The Sociological Quarterly*, 18, Winter.
Curran, J., Jarvis, R., Blackburn, R.A. and Black, S. (1993) Networks and Small Firms: Constructs, Methodological Strategies and Some Findings, *International Small Business Journal*, Jan–Mar, 11, 2.
Dodgson, M. (1993) *Technological Collaboration in Industry. Strategy, Policy and Internationalisation in Innovation*, Routledge, London.
Fletcher, D. (1993) *Small Firms Strategic Alliances and Value Adding Networks. A Critical Review*, paper presented at RENT VII Budapest, November.
Granovetter, M. (1973) The Strength of Weak Ties, *American Journal of Sociology*, 78.
Gustafsson, J., Henriksen, L.B. and Larsson, R. (1991) *Udvikling af strategiske alliancer, joint venture og netværk*, DIOS & AUC, Ålborg.
Hamel, G. and Prahalad, C.K. (1994) *Competing for the Future*, Harvard Business School Press, Boston, MA.

Industri- og Handelsstyrelsen (1990) *Virksomhedsnetværk – Juridiske, skattemæssige og regnskabsmæssige aspektiver ved netværk*, Kbh. 1990.

Jarillo, J.C. and Stevenson, H. (1991) Cooperative Strategies: The Payoffs and the Pitfalls, *Long Range Planning*, 24, 1, pp. 64–70.

Johannisson, B. (1986) *Beyond Process and Structure – Social Exchange Netwocouraging Entrepreneurship – Progress Report from Field Research into Stagnating Swedish Communities*, Babson College workshop on encouraging entrepreneurship internationally. Wellesley, MA, April.

Johanson, J. and Matson, L.G. (1988) 'Interorganisational Relations in Industrial Systems – A Network Approach' in Hood, M. and Vahlne, A. (eds.) *Strategies in Global Competition*, Croom Helm, London.

Lorange, P. and Roos, J. (1992) *Strategic Alliances: Formation, Implementation and Evolution*, Blackwell, London.

Munksgaard, T. (1995) Når kemien passer . . ., Master Thesis, Copenhagen Business School.

Mønsted, M. (1995) Processes and Structures of Networks: Reflections on Methodology, *Entrepreneurship and Regional Development*, 3, 3, July–October.

Normann, R. (1984) *Service Management*, Copenhagen.

Powell, W.M. (1987) Neither Market nor Hierarchy: Network Forms of Organisation, *Research in Organisational Behaviour*, 12, pp. 295–336.

Powell, W.W. and Brantley, P. (1992) Competitive Cooperation in Biotechnology: Learning through Networks? in Nohria, N. and Eccles, R.G. (eds.) *Networks and Organisations*, Harvard Business School Press, Boston, MA.

Powell, W.M., Koput, K.W. and Smith-Doerr, L. (1994) *Technological Change and the Locus of Innovation: Networks of Learning in Biotechnology*, paper presented at the Academy of Management, Dallas.

Powell, W.W. and Smith-Doerr, L. (1994) Networks and Economic Life in Smelser, N.J. and Swedberg, R. (eds.) *The Handbook of Economic Sociology*, pp. 386–402, Princeton University Press, Princeton.

Pyke, F. (1988) Cooperative Practices among Small and Medium Sized Establishments, *Work Employment and Society*, 2, 3, pp. 352–65.

Stinchcombe, A.L. (1989) A Outsider's View of Network Analyses of Power in Perrucci, R. and Potter, H.R. (eds.) *Networks of Power: Organisational Actors at the National, Corporate, and Community Levels*, De Gruyter, New York.

Schumpeter, J. (1934) *The Theory of Economic Development*, (reprint 1978), Oxford University Press.

Töennies, F. (1957) *Community and Society*, East Lansing, Mich.

Williamsson, O.E. (1975) *Markets and Hierarchies: Analysis and Antitrust Implications*, Free Press, New York.

CHAPTER 10

How Entrepreneurial Networks Can Succeed: Cases from the Region of Twente

ROSALINDE KLEIN WOOLTHUIS, DENNIS SCHIPPER AND MARTIN STOR

INTRODUCTION

In general firms co-operate because they are forced to do so as a result of rapid developments in their environments (e.g. the pace of technological change), either because they seek to exploit a market opportunity together (i.e. entrepreneurial incentive), or because they are encouraged to do so through external measures (e.g. government subsidies, policy measures etc.). However, the process of starting and maintaining an inter-organisational relationship is difficult and time consuming. Sometimes co-operation leads to success, although very often co-operation is a complicated process that is prematurely terminated. Interested parties must be brought together, while a trusting atmosphere has to be established. Moreover, an organisational structure must be established and commitments should be made and executed in order to be able to reach the stated goals of an agreement. Also during this process, the inter-firm relationship is vulnerable to the various distortions of an economic, technological and social nature (Larson, 1991; 1992; Ring and van de Ven, 1992; 1994).

In the region of Twente in the Netherlands there is considerable experience with numerous forms of co-operation. Through Innovation Centres, the Dutch government tries to stimulate innovation and co-operation among small and medium sized enterprises (SMEs), especially in the less developed regions such as Twente which is a former textile region. In addition, the University of Twente is actively involved in these kinds of initiatives. The University functions as a sort of consulting institution for co-operating parties, and is active in studying the various forms of co-operation in the region. From this involvement, both the Innovation Centre and the University have discovered how co-operation can be established, how co-operative projects evolve over time, and which factors contribute to, or hinder, the success of such co-operations.

The aim of this paper is to explain this experience. First we will describe the way in which the Innovation Centre tries to stimulate co-operation between firms. Three methods to enhance co-operation are discussed and 3 examples are given to illustrate the functioning of these methods. However, to get a deeper understanding of the process of co-operation, an in-depth longitudinal case study is described in the second part of the paper. From this case, the development of inter-personal and inter-firm relations can be traced and their influence on the success and course of the joint innovation projects is assessed. One of the participants of the network, the design office Demcon Twente b.v., explains the experience of initiating and joining co-operation networks. From all these experiences, lessons can be derived for other co-operation projects and insights can be gained into the process of co-operation. Out of these lessons we have composed '10 commandments of inter-firm co-operation', which we believe reflect some major pitfalls and success factors experienced during co-operation. With these 'rules of thumb' in mind, co-operation projects should be easier to overview and 'best practice' strategies can be found for the management of co-operation projects.

POTENTIAL DIFFICULTIES WITH THE CO-OPERATION PROCESS

Experience has shown that inter-firm co-operation plays a role of increased importance when companies innovate. This is not surprising if it is noted that at this moment there is a form of hyper-competition taking place, in which only the best firms will survive. In the development of a new product-market combination, various skills are required. First, market signals have to be picked up and well interpreted, while there is an additional need for design and manufacturing skills. There is little chance that a single small firm has all these skills at its disposal of the required level. Therefore, this prompts the firm to engage in some form of co-operation. However, co-operation between SMEs often fails. Problems that were encountered by the Innovation Centre in establishing co-operation between firms were:

- different orientations of SMEs (like marketing, production or more research oriented) made co-operation difficult,
- co-operation proved to be a difficult and time consuming process because of the complexity of the projects in some sectors since developments could take up to 3 years,
- for successful product developments all partners had to master the state of the art technologies, since a lack of knowledge of one party could spoil the project,
- because of the complex character of the products – which often required high investments and offered relatively small (niche) markets – risk was high.

Different causes can be identified for these difficulties in establishing successful co-operation. One of the major underlying causes of difficult co-operation, which is specific for SMEs, is that small companies are often managed by owner-managers (Aldrich and Zimmer, 1985; Aldrich et al., 1987). A quite common reason for founders to become self-employed is that they find it difficult to collaborate. Entrepreneurs often show a strong sense of independence and high trust in their own capabilities and insights. Usually they were successful by following their own judgements, which strengthened their belief in independent strategy making. Therefore they often found it difficult to listen to others when co-operation was attempted. But there are more reasons why co-operation might prove to be difficult.

Communication has proven to be one of the most important and difficult topics when parties start their co-operation. People becoming involved in a network have to learn to interpret the verbal and non-verbal communication of the other parties involved. The way people communicate can differ by company, by region or by the country from which people originate. Thus, the meaning of language can differ between people and situations. People involved in co-operations need to learn to cope with these differences in interpretation, especially because, through communication between people, bonds of trust can develop. If people perceive another party to be comprehensible in their communications, they will find it easier to attach the correct meaning to the communication. If such parties behave according to the perceived meaning they will be considered understandable and predictable. This makes it easier to trust this party. Such trust, in turn, is essential to doing business together, because trust decreases the fear of the opportunistic behaviour of others (Zand, 1972; Nooteboom, 1995; Gulati, 1995).

In order to be able to work on such a trust basis, a number of 'rules of thumb' have to be kept in mind. Major difficulties can, for example, occur if competitors attempt co-operation. The fear of opportunistic behaviour may be great because the parties are often competing for the same share of the market, the same customer, or the same technological leap forward. This makes it difficult for the parties to trust each other. They will constantly evaluate the contribution of the other party, and the benefits they derive from the joint project. There will be a very high standard set for the perceived extra value that the competitor derives from the co-operation. However, parties will probably react in a negative way if they perceive this return to be higher than their own benefits. Therefore conflicts will easily arise and trust is likely to be very vulnerable. It is thus advisable not to co-operate with competitors unless the co-operative project is in an early stage of the innovation process and the information exchanged, and the knowledge developed, is not of a competitive nature.

THE NETWORK STRUCTURES OF THE INNOVATION CENTRE

The Innovation Centre Network of The Netherlands (ICNN) is well aware of the difficulties that can arise during the process of co-operation. However, it is, at the same time, aware of the opportunities co-operation can bring to SMEs during the process of innovation. Because smaller firms are limited in their material and immaterial resources, the benefits of bringing together complementary parties can be substantial (Brockhoff, 1992). Therefore the Innovation Centre Network supports 3 different types of co-operational structures between firms in which they may try to 'coach' the co-operation into a successful outcome. The Innovation Centre seeks to attain a leading position in establishing and facilitating these types of co-operation. Their experience of consulting SMEs in the field of innovation, their national network, and their insights on, and contacts with, Dutch SMEs, make them a unique knowledge-centre in the Netherlands.

The 3 forms of co-operation which the Innovation Centre supports and coaches are named Meshwork, Business Support Circles and Knowledge-Intensive Industry Clustering. These forms have different structures and dynamics as follows (also graphically depicted in Figure 10.1):

- Meshwork is a systematic approach to acquiring a proper combination of companies that can jointly develop new products,
- Business Support Circles (BSC) are not based on such a systematic approach. In this case the ICNN brings companies together which can possibly lead to co-operation,
- Knowledge-Intensive Industry Clustering is a type of co-operation in which SMEs act as a joint main supplier of large customers.

The different forms of co-operation serve different goals and each has its own advantages and possible pitfalls. Next we will give a short description of the different organisation forms and illustrate them with a case study example.

Meshwork

The main objective of the Meshwork project is to increase the number of market opportunities for SMEs through greater co-operation. In this type of project, new co-operation structures between SMEs are created by first looking for a market opportunity, and then by matching SMEs that have the joint core skills needed to exploit a

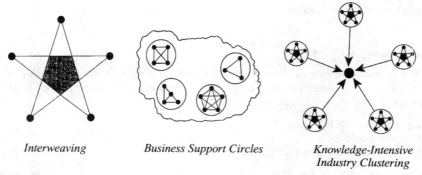

Figure 10.1
Characterisation of the 3 types of co-operation

market opportunity. To do this a 'Meshwork Method' has been developed. This method first conducts a 'Meshwork Quickscan' for companies that are interested in co-operation with regard to this type of project. The objective is to get a global impression of the company, and of its capabilities, which might make it a suitable company for co-operation over joint product development. This Quickscan takes about 2 hours and is executed during an interview between the manager of the SME and an Innovation Centre consultant. It results in a short report (one page of A4) in which the culture of the SME, the dominant thinking routine and its innovation potential are visualised.

The second step concerns the 'Meshwork Core Skill Scan'. The objective of this step is to obtain an insight into the core skills that the SME can offer to a potential (Meshwork) partner. Again a strong visual tool is used to acquire the information needed. Normally the SME's management participates in this session. The duration of the scan is approximately 4 hours. This session results in a report visualising the capabilities and the core skills of the SME.

After this step a 'Meshwork Market Opportunity Workshop' is organised to try to bring the companies' capabilities and potential market opportunities together. This workshop has a creative character. After defining the strategic options, market opportunities are then generated. This session also takes about 4 hours. Besides this initiative, the data acquired in the Meshwork Core Skill Scan, and the data from the Market Opportunity Workshop, are stored in the Darwin data bank. This computer data bank thus contains all the information on potential meshwork partners. The Innovation Centre consultants are the only persons that have free access to such data. With the help of this data bank they actively search for new 'matches' of companies to exploit market opportunities.

A case example – the Tilburg bicycle parking
The city of Tilburg saw an opportunity to introduce fully automated underground bicycle parking. A construction firm recognised this possibility and carried out a market survey. This survey indicated that a market existed for the idea, if a machine could be developed which was both technically and economically feasible. With the assistance of the Innovation Centre, 5 companies were found with complementary skills and resources. Each participant in this Meshwork network co-operated at their own risk in the design of such a machine. The project partners negotiated and designed a fully automated machine with a capacity of hundreds of bicycles. A design offering was made to the city of Tilburg. After being delivered, the bicycle is automatically parked by a robot. This robot is also responsible for safely returning the bicycle. The parties

worked together with other partners in the field of process control, mechanical engineering and electrical installations.

Business Support Circles

Entrepreneurs within the same sector are not merely competitors, since they have discovered that the exchange of knowledge and information can bring profit for them all. Examples of such co-operation are:

- shared investment in R&D,
- shared export projects,
- shared development of product standards.

In business supporting circles, entrepreneurs are involved in an informal way. In the beginning, the development of personal relationships is the main objective. A goal of developing inter-personal contact is the establishment of bonds of trust between entrepreneurs. When trust relationships have been achieved in the personal sphere, then entrepreneurs may explore the possibilities for inter-firm relationships. These business supporting circles can also be organised around a technology. Examples of this are projects which the Innovation Centre promotes in medical technology, materials technology and mechatronics. The ICNN can facilitate these business circles by finding information, defining the project, providing funding and facilitating coaching of the process.

A case example – co-operation in agriculture

There are a lot of examples of these business supporting circles. An interesting case is that of farmers who wanted to develop new machines in order to decrease labour costs. The Innovation Centre in The Hague organised a co-operative project between the farmers, suppliers, and an insurance company. All the participants invested in:

- the improvement of the internal transport of tomatoes,
- the development of a tomato harvesting robot,
- the development of a radish harvesting machine.

Essential in all these projects was the transfer of knowledge between industrial sectors. For example, there is a lot of robot technology available in the Netherlands, and this technology is now being used in the farm machinery sector.

Knowledge-Intensive Industry Clustering (KIC)

The KIC projects aim at increasing the competitive strength of the sub-contracting industry. In this project, clusters of complementary sub-contract suppliers collaborate with a large company in the development of new products. This results in a shorter development time of the new product and in a higher quality.

In the first stage of these projects the outsourcing company and the Innovation Centre define components of a finished product that are suitable for outsourcing. Then an inventory of potential suppliers is made who are able to co-operate in the design and production of the components. From this group, a cluster of complementary suppliers is selected. If needed, knowledge is supplemented by external knowledge sources. Assisted by the outsourcing company and the IC-consultant, the cluster can then start the development. The conclusion of this stage will be the building of the prototype. One of the suppliers is in charge of the development of the component. In the last stage of the project, the production of the component begins.

A case example – OCE Nederland b.v.
OCE develops new products such as colour copiers and colour printers using state-of-the-art digital technology. OCE expects to introduce new copiers in 1997. For this purpose, the company has co-operated with about 45 clusters in the south-east of the Netherlands. The products are divided into component parts modules (e.g. the control panel of a machine). Each cluster has the task of developing a module. In the clusters, R&D, engineering and production are brought together. The early participation of mechanical engineers, electricians, injection moulding experts and process engineers has led to a better product and a shorter product development time. According to the R&D manager of OCE the Knowledge-Intensive Industry Clustering project is of great relevance. He thinks that it is the start of a new strategic direction that his company has to follow to be able to remain competitive in the industry.

From the projects described above, the Innovation Centre has gained considerable experience in inter-firm co-operation. To obtain more in-depth insight of how co-operation can be established, and of how relationships can develop over time, a time path analysis of co-operation in innovation is given in the following TIMP case. TIMP is fully titled the Twente Initiative on Medical Products and is aimed at introducing new, innovative medical products with shorter lead-times and lower costs. Due to the multidisciplinary origin of the parties involved, the technological risk of the projects was assumed to be lower. Further, such co-operation was assumed to lead to an expansion of the current activities in the region. We will describe the case to illustrate how co-operation can come into existence, what organisation structure can be chosen, and how joint innovation projects can take place.

THE TIMP CASE

The TIMP foundation is an entrepreneurial network in the region of Twente. The co-operation achieved by this body is aimed at product development in the medical care sector. Medical technologies are developed to serve the market segments of home care and rehabilitation. Due to the proportional increase of an ageing population, increasing market demand can be expected. To exploit this market opportunity, a combination of technological capabilities, market knowledge and marketing skills is needed. Because of the industries' multidisciplinary and specialist character, theoretical knowledge must be combined, and must be translated into practice, in order to develop commercially successful products. In Twente, there were different companies and institutions working in the medical field. Extensive theoretical and practical knowledge was present at the university and at a number of technology-based firms. Thus, a combination of these competencies might help the firms to compete, which would also contribute to regional development. In the Innovation Network terminology, TIMP can be understood as a Business Support Circle.

The initiative started with the realisation that a number of firms could jointly achieve innovation, which might be difficult to achieve individually. This awareness resulted in a simultaneous initiative of some technically based companies to approach the Overijsselse Ontwikkelings Maatschappij n.v. (OOM) (an innovation subsidising institution for the region) to bring together a number of companies with complementary expertise, in order to make joint innovation projects possible. This initiative could be supported with a subsidy from the European Fund for Regional Development (i.e. EFRO). Around February 1995, 12 parties gathered around the table and negotiations started on the co-operation agreement to be achieved.

The early orientation phase included a presentation of the parties interested in the co-operation. Entrepreneurs could, in this way, form an impression of each other in order to determine whether they wanted to participate in the network. Additional partners were sought where additional capabilities were needed. Also, in order to ensure a strong bonding of the network, a good 'fit' between partners was crucial. The fit between companies refers to strategic goals and business cultures, together with similar image and quality factors.

The composition of TIMP

The TIMP group is composed of parties which are mostly complementary in nature, although, some partly compete with each other. The 12 include 2 market-oriented firms, 3 government institutions, and 6 supplier companies. The market-oriented firms operate in different market segments of the medical sector, but have joined the co-operation because they are interested in obtaining new innovative products for their markets. The 3 government institutions involved are the Regional Development Agency (OOM), the IC-Enschede (Innovation Centre) and the University of Twente. Their motive in being involved in this co-operation is, on the one hand, to try to support the development of the region, and on the other, to obtain more experience in the process of facilitating co-operative initiatives. The University of Twente plays a special role in the sense that it wishes to apply theoretical knowledge in practice involving technological and managerial expertise, while also wanting to learn from practical experience.

The 6 supplying companies include 3 design-companies, an international trading company, and 2 companies specialising in the ergonomic aspects of medical products. The design companies have different core competencies including mechatronics, electronics and industrial design. However, they also possess each other's capabilities, which tends to make them partly competitors. The international trading company involved is interested in the possible outcome of the innovation projects which might be internationally marketable as products, patents or licensing agreements. This also offers the design and production companies the potential to expand their markets beyond their regional or national markets. The last 2 companies are a physiotherapist and a company specialising in ergonomic design, which are able to judge the quality and user-friendliness of the developed products.

The companies are all small, with the largest company being not more than 30 employees. Most companies are young, ranging from 1–10 years in age, and founded by university graduates with a technical background. Thus, the network mainly consists of entrepreneurially-based firms. According to Larson (1991) the entrepreneurial firm (defined as adaptive and innovative) is said to especially benefit from external relationships, since it offers a leveraging opportunity and the network structure enhances the adaptability of the firm to its environment. In all cases, these entrepreneurs represent their company in the meetings of the network.

The structural development of the co-operation agreement

To find a proper legal and organisational form for the co-operation, the regional institutions took on the role of consultant, to help to find an organisational structure, and to promote the still vulnerable process of co-operation. This included chairmanship at meetings, encouraging and guiding the first contacts between parties, and establishing contacts with external parties. These contacts were needed in order to apply for the EFRO subsidy, and to provide the notary services required to establish the legal form of the co-operation. At first they aimed to have a 2 hour meeting every 3 to 4 weeks, but later in the co-operation this proved to be too often. The entrepreneurs wanted to get on with the business, and were less interested in the administrative aspects of the agreement. Soon the meetings were scheduled to occur every 2 to 3 months.

Eventually, a memorandum of association was signed by all participants. The legal structure chosen for TIMP was a Foundation, which implies that decisions are made democratically, and that the board is chosen by its members. Daily affairs are taken care of by a 3 man board. The TIMP organisation can be envisaged as a sort of 'umbrella' organisation in which all parties are represented. However, under this 'umbrella' different initiatives can arise. Product ideas are brought to the notice of the general organisation and then possibilities are reviewed as to which partners can best co-operate in developing the product. When the innovation is ready for the prototype phase it is introduced into the network. The design is ready, the market has been surveyed, and so the product is ready for further development, production and sale into the market. Thus under the main umbrella, smaller clusters of firms are configured which co-operate in the development of a product idea. These clusters mostly include a market partner who is the product champion, who provides the product idea, and bears both the investment and risk. The co-operating parties play a supplying role in developing and advising on the development of the product. Each partner does the part he can do best, such as the provision of electronics, design or ergonomics skills. The value added of the network is that all partners review the product ideas critically and give their suggestions for improvement from their specific field of knowledge. Furthermore, informal channels of communication are formed in the network for consulting purposes, although they are often triggered at a very early stage of the more formal networks development.

Since the product development costs are subsidised to the value of 50% by EFRO (a subsidy of the EC for regional development) it is very important that subsidies are spread fairly over the group. Most subsidy is obtained by those parties that invest most (and take most risk) in the new product development. In the TIMP case these are the market-orientated parties. This implies that they, in turn, have to take care of a more or less even distribution of work in the group to prevent conflict and obstruction of the process of co-operation. Perceptions of an unfair distribution of work and subsidy will lead the aggrieved parties to become dissatisfied with the benefit of the co-operation, which can lead to conflict. To oversee the fair distribution of work and subsidy, external parties are employed who co-evaluate the submitted product proposals. For this purpose, an external board is established including a representative of the Dutch Ministry of Economic Affairs, a facility manager of a regional hospital, and a director of medical home-care institute.

In the execution phase, the companies become more strategically interdependent because the success of the champion's product is dependent upon the capabilities of the sub-contracted parties. The sub-contracted parties, in turn, are dependent on the turnover that can be achieved by being hired by the product champion. Satisfactory execution of the job is important for repeat contracts. The importance of the continuity of a relationship is thus present for both the sub-contracting champion and the executing sub-contractor party. The sub-contractor party will perform reliably to ensure future assignments, while the champion benefits from this because he has less worry about the quality of the work performed, or the possibility of opportunistic behaviour. Continuity of the relationship is thus a result, but it is also a governance mechanism of the inter-organisational relationships. This effect is often referred to by game theorists (Neumann and Morgenstern, 1944; Axelrod, 1984).

The relational development of the co-operation

The start of the co-operation was difficult and time consuming. During the first couple of months, there were problems of the interpretation of others' behaviour and problems with communicating due to personal and cultural differences. Therefore it was important that the initiators took on a guiding role. They helped to find a proper legal and organisational form and aided the still vulnerable process of co-operation. This enabled the participants to concentrate on getting to know each other, and to look for market or product development opportunities. Furthermore it prevented early conflict since the initiating role, which could be interpreted as a dominant role, was taken by an independent party which has no financial interest.

Another important aspect contributing towards the positive development of the relationships was that most parties were familiar with each other before they engaged in the TIMP co-operation. They had previously met during ordinary business transactions or knew each other from other non-business activities. In most studies of inter-organisational relationships, the importance of familiarity with business partners is stressed (Hakansson, 1987; Larson, 1991), because familiarity enables parties to better evaluate the capability (or reliability) and the willingness (or trustworthiness) of the other party. This can speed up the initiation phase, while making commitments to the co-operation easier, because parties have less fear of opportunistic behaviour.

The familiarity of TIMP partners helped them to evaluate the potential of the group, and to estimate the possible difficulties that could occur with the group's composition. An important additional factor prompting the parties to engage in a co-operation was the trust most parties also had in the initiators, the OOM and the IC. Because they trusted *their* ability to compose a promising network, they also trusted the parties that they introduced into the network. Trust can thus be transmitted through parties that have already obtained a 'to be trusted' status. This was also the case with the business partners. Most companies considered their own familiarity with the other party *or* someone else's knowledge of that party of great importance. Parties that were new to all network members were considered with great care, and members were reluctant to exchange information with these parties, of whom it was not completely clear which role they would fulfil or what was their actual motive in joining the co-operation.

For this reason secrecy of information was also very important to the participants since an unwanted transfer of information on new products could mean a loss of competitive advantage in the future. As soon as the first agreements were made, a pledge of secrecy concerning matters of development, production, marketing and knowledge/technology transfer had to be signed by each of the network members. This document would be valid to 5 years after the end of the co-operation agreement. However, the parties realised that the *strict* protection of knowledge and know-how was impossible despite a signed contract. Trust served as alternative mechanism which removed the fear of unwanted knowledge transfer.

This reliance on trust is an aspect of the co-operation that only gradually developed. As mentioned before, trust was partly present because of the familiarity of partners, partly because trust was transferred through other parties, but a true trust basis still had to develop through the experience of firm-interaction during innovation projects. A contributing factor to the development of trust in the TIMP case was that they had a shared culture because they were all from the same region. This gave them a feeling of solidarity. They could also speak to each other using their own accent. The business expertise of the partners may also play an important role in easing inter-firm communication because partners from the same technical or educational background are able 'to speak the same language'. Finally, the TIMP group as a whole could take on an organisational culture which implied shared norms and values which made it easier to predict partner behaviour, and to trust each other.

In the TIMP case the regional aspect is quite important. The EU subsidy is aimed at regional development which seeks to promote the mutual support between participants of TIMP group. Furthermore, the technical knowledge of the participants coincides for the most part, since many of the companies have a strong technological orientation, and almost all entrepreneurs gained their education at the technical university of Twente.

These similarities enhanced a feeling of trust, and thereby also contributed to open communication between the parties. The openness of a shared culture is crucial for the co-operation to fulfil its goal and the development of new products and technologies. To be able to innovate, a free flow of information and ideas is necessary, in order to produce high quality creative products. If a defensive climate arises among parties, information is not exchanged, and commitment is difficult to obtain and maintain.

Experiences of a TIMP member: Demcon Twente b.v.

Demcon is a small entrepreneurial firm which specialises in mechatronic design. Demcon was one of the first parties interested in the founding of TIMP. The major reasons for Demcon to initiate TIMP and to invest time and money in this network was:

- the opportunity to create business by selling its mechatronics design capacity to the market-orientated members of TIMP,
- the opportunity to become familiar with the field of medical product development,
- to gain experience of co-operation with the other design companies in Twente in the field of medical product development,
- to become known as a company which is willing to co-operate, and which can be trusted.

After a long period of preparation and getting to know the other parties, TIMP has become very successful for Demcon. Demcon is involved in at least 5 projects, in which it plays an important role. In the TIMP projects Demcon functions as a supplier of mechatronics design knowledge, but also acts as an initiator of new product developments when they feel that a new product has a clear market opportunity. But membership of TIMP has provided Demcon with even more benefits. Because the relationship between the TIMP member firms has improved (i.e. the entrepreneurs have got to know each other's capabilities and have gained trust in each other), room was created for deeper forms of co-operation. Between the 3 design companies involved, co-operation arose, which was not restricted to TIMP projects. The aim of this co-operation is to create more business opportunities by using the combined capacity of the 3 companies. This co-operation is only possible because the 3 companies have different core competencies. Another important reason of this unique co-operation, and probably the most important one, is the good relationship between the directors of the companies which is based on trust. In the near future Demcon would like to participate in, and initiate, new networks that are created to exploit promising market opportunities. The experiences gained will possibly result in a more efficient preparation phase for the networks and probably in more accurate selection of product ideas and network partners.

CONCLUSION

The aim of this paper was to report on the experience gained by supporting and examining inter-firm co-operation in the region of Twente. The experience of the Innovation Centre Network the Netherlands and the University of Twente led to the composition of a list of 10 rules of thumb which we feel are very important to keep in mind when an inter-firm relationship is being established and maintained. These rules are presented in Figure 10.2.

	The 10 Commandments for Setting Up and Maintaining Successful Inter-Firm Relationships
1. Show respect for the market!	Only if there is reasonable market potential has co-operation a chance of survival. The market often determines the composition of the network and the jobs that should be performed.
2. Know and respect each other's core skills	In inter-firm relations there have to be complementary core skills. If one partner supplies the design, another takes care of the production of metal parts, while the next produces the plastic parts and takes care of the assembly, and the last one knows the right market channels and is responsible for sales and marketing.
3. Make sure there is clear-cut communication	Co-operation is between people. This implies that good open communication is of vital importance. Each partner has his own way of thinking and world of experience. Identify and recognise each other's preferred way of thinking to prevent conflict.
4. Know your own and each other's interests	At each stage of the co-operation it should be clear to everyone what his own interests and those of the others are. Openness is necessary, hidden agendas are forbidden!
5. Make sure there is commitment!	A co-operation project consists of different stages. In each stage the partners should show commitment. Commitment is shown through action not words.
6. Co-operation should never lead to an increase in costs	The market is the judge. If an outsider evidently does better than one of the partners it should be possible to involve this outsider in the project. The network partners should be given the right of entry to the group.
7. Co-operating is a long-term project	Sometimes things are not as they should be. In such a case the partners are expected to invest in the relationship to put things right. Joint innovation is often an unpredictable, and especially laborious trajectory.
8. Make arrangements beforehand on how to share the profits	Do not use complicated contracts, but make clear, simple arrangements that leave room for flexibility where necessary and are acceptable to everyone. It should include rules on how to share profits.
9. Make sure there is good co-ordination	There has to be a co-ordinator during the different stages of a project. This could involve the hiring of a partner or an outside expert. Outside partners are advisable because they can independently supervise the process.
10. Co-operate in order to inspire one another!	Learning and stimulating each other is not only fun but will also enhance the process of co-operation and the products that are developed.

Source: M. Stor and .W. Rutenfrans, Innovation Centre Overijssel

Figure 10.2
The 10 commandments of inter-firm relationships

From the in-depth longitudinal case study of the TIMP case, the development of inter-personal and inter-firm relations can be traced. Success in the TIMP case seems to stem from different sources. First there is the joint interest to exploit market opportunities that members cannot achieve alone, by using the complementary technological capabilities and other resources available within the TIMP partner group. The TIMP organisation offers members a possibility to enhance their potential in co-operation with others. It is important here to note that the customer and supplier roles were very clearly separable because the market-orientated members who manufactured the final product, in most cases, had a sub-contracting function. This made possible a clear distribution of tasks, responsibilities, risks and benefits. A contributing factor in this respect was also the subsidy that could be obtained, which meant that more innovation projects could be realised and an increase in turnover of a number of partners could be achieved. These represented the inter-firm, economic incentives.

On the other hand, inter-personal relations played an important role. Part of the success of the network can be explained by the common interests of the partners. They are not only from the same cultural background, but they also speak the same language because they have the same educational background (engineers), the same professional background (entrepreneurs) and have similar company cultures (professional engineering organisations). This enhances the ease of communication and the development of trust in a co-operation.

NOTES

The Innovation Centre Network Netherlands

Martin Stor is consultant for the Innovation Centre in Enschede. The task of the Innovation Centre Network Netherlands (ICNN) is making new technology accessible and applicable for small and medium-sized enterprises. The ICNN employs about 300 staff. The most important target group are industrial SMEs with no more than 50 employees. ICNN's most important financier is the Dutch Ministry of Economic Affairs. Previously support was aimed at actively transferring new technologies to SMEs. Gradually, more and more attention was paid to other forms of innovation, such as developing new product-market combinations and inter-firm co-operation in innovation.

The University of Twente

Rosalinde Klein Woolthuis is a Ph.D student at the School of Management Studies of the University of Twente. In her Ph.D project she examines how inter-organisational relationships develop over time and how this process is influenced by relational and economic factors. Her research is part of the research programme on entrepreneurial networks, supervised by Prof. dr ir. W.E. During. In this programme the university aims to obtain understanding of the functioning of inter-firm relations between small entrepreneurial firms and how this process can be enhanced. The university plays an important role in regional initiatives to use inter-organisational relationships to innovate.

Demcon Twente b.v. – Hengelo

Demcon Twente b.v. is a design company in the field of mechatronics design. In its 3 year of existence the company has obtained a reputation in designing innovative products in different market areas. Besides a multidisciplinary team of 10 designers (mechanics, electronics and software) Demcon has its own mechanical workshop with experienced staff members.

REFERENCES

Aldrich, H. and Zimmer, C. (1985) Entrepreneurship Through Social Networks, in Smilor, R. and Sexton, D. (eds.) *Entrepreneurship*, Ballinger, New York.

Aldrich, H., Rosen, B. and Woodward, W. (1987) The Impact of Social Networks on Business Foundings and Profit: A Longitudinal Study, Babson College, Wellesley, MA.

Axelrod, R. (1984) *The Evolution of Co-operation*, Basic Books, New York.

Brockhoff, K. (1992) R&D co-operation between firms – A perceived transaction costs perspective, *Management Science*, 38, 4.

Gulati, R. (1995) Does familiarity breed trust? The implications of repeated ties for contractual choice in alliances, *Academy of Management Journal*, 38, 1, pp. 85–112.

Håkansson, H. (ed.) (1987) *Industrial Technological Development: A Networks Approach*, Croom Helm, London.

Larson, A. (1991) Partner Networks: Leveraging External Ties to Improve Entrepreneurial Performance, *Journal of Business Venturing*, 6, pp. 173–88.

Larson, A. (1992) Network Dyads in Entrepreneurial Settings: A Study of the Governance of Exchange Relationships, *Administrative Science Quarterly*, 37, pp. 76–104.

Neumann, J. von and Morgenstern, O. (1944) *Theory of Games and Economic Behaviour*, Princeton University Press, Princeton.

Nooteboom, B. (1995) Trust, Opportunism and Governance, Research report, RUG.

Ring, P.S. and van de Ven, A.H. (1992) Structuring co-operative relationships between organisations, *Strategic Management Journal*, 13, pp. 483–98.

Ring, P.S. and van de Ven, A.H. (1994) Developmental processes of co-operative interorganisational relationships, *Academy of Management Review*, 19, 1, pp. 90–118.

Zand, D.E. (1972) Trust and Managerial Problem Solving, *Administrative Science Quarterly*.

CHAPTER 11

Trust and Management: As Applied to Innovative Small Companies

ETIENNE KRIEGER

INTRODUCTION

Since the beginning of the nineties the notion of trust has increasingly been evoked, and sometimes invoked, mainly because it seems to be sadly lacking in several areas of human activity; such as in economics, politics and society in general. When 'everything is right', trust is so obvious that one forgets its contribution to human interactions. But when the economic situation turns bad, when distrust creeps in and hinders initiative, innovation and growth, the retrieval of trust seems to be the only way to cut the Gordian knot of strictly calculative interactions.

This observation applies at several levels: interorganisational (relationships between banks and firms, between customers and suppliers, etc.), intra-organisational (specific management methods, decentralised vs. centralised structures), and inter-individual relationships.

Having observed the birth of several innovative companies and their respective evolutions in terms of fast growth, stagnation and, sometimes, failure, we very often notice that the destiny of these companies cannot be explained by strictly technical or economic parameters. Are the real factors of development to be found elsewhere? Can we analyse, or at least understand, the genesis and the success of new firms in the light of *trust*?

This paper will clarify the concept of trust through the observation of the growth and the development of new ventures. It will define its effects and its limits and examine, with the help of this concept, several theories and various practices in the field of strategic management[1].

TRUST, INNOVATION AND NEW VENTURES: SOME DEFINITIONS

Most authors seem to be reluctant to give a definition of trust[2], preferring instead to describe the outcomes, the factors that emerge from the various forms of trust. Nevertheless, the following definitions of trust are suggested below:

- 'To have belief or confidence in the honesty, goodness, skill or safety of (a person, organisation or thing)' (Cambridge International Dictionary, 1995),
- 'Feeling that enables reliance on someone or on something' (Servet, 1994),

- 'Action connected to interest but with an assumed risk, in a non probabilisable universe' (Coriat and Guennif, 1996 – definition of 'pure trust'),
- 'Assumed risk, driven by human intelligence that only relies on itself' (Peyrefitte, 1995),
- 'Trust is a disposition to take a risk exposing oneself to other's opportunism' (Van Wijk, 1996),
- 'The extent to which a person is confident in, and willing to act on the basis of, the words, actions, and decisions of another' (McAllister, 1995),
- 'Assumption that, in case of uncertainty, the other party will act – even in unforeseen circumstances – according to rules of behavior that we find acceptable' (Bidault and Jarillo, 1995).

We shall adopt the following definition of trust: positive anticipation associated with an assumed risk. This definition includes the notions of responsibility, of choice, and a mix of rationality and non-instrumental expectations.

Furthermore, we shall retain a definition of new ventures consistent with the approach of French public institutions: their framework of intervention for 'business creation' ranges from -1 to +3 years. As a consequence:

- the company itself may not necessarily have started, but its incorporation is imminent,
- the company is considered to be a new venture until the end of its third year.

The classification of high tech companies suggested by Albert and Mougenot (1988) is the following:

- *advanced technology companies*: between science and industry, these companies develop and/or exploit new technologies in different applications for several markets,
- *innovative firms*, which launch new products or processes with a technological content, derived from classical or advanced technologies. These companies have some characteristics in advanced technology companies but their know-how is focused on the application of technology into the served market,
- *innovative firms*, which are traditional firms launching new products or services or using new processes, involving marketing or management methods without necessarily having a technological content.

We chose to study a population of 'so-called' innovative companies, whose characteristics include the commercial launching of new products or services, or the use of new methods of manufacturing, marketing and management.

MANAGERIAL LITERATURE AND THE PRINCIPAL TRENDS OF RESEARCH

If one considers trust as a relationship, the analysis of the managerial literature allows us to suggest the following model (Figure 11.1):

Several conceptions of trust can be found in the managerial literature, partly because trust can be expressed through different modes:

- It can for example be limited to a given deadline or purpose. But trust can also be more unconditional and more diffuse. This will then be a form of open trust[3], which is not *focused* on a specific goal,
- It can be analysed in a *hierarchical relationship* involving vertical trust, or in a relationship between peers[4] representing horizontal trust,

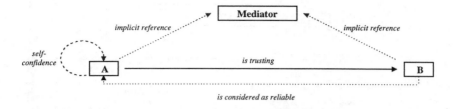

Figure 11.1
Describing the relation of trust between 2 parties

- Although trust is a dual notion[5], it can play a role at several levels of an organisation such as interpersonal (individual), inter-departmental (group) or interorganisational (company)[6],
- Finally, several authors make a distinction between *affect-based trust* and *cognition-based trust*[7]; or between *moral* and *technical trust*[8].

We can therefore propose the following summary table indicating the different modes of trust:

Openness	*Focused* trust vs. *open* trust
Power relationship	*Vertical* trust vs. *horizontal* trust
Level	*Individual*, *group* or *organisation*
Nature	*Moral* trust vs. *technical* trust

In many books or articles, it appears clearly that trust spreads over the entire value chain of a given firm, or a given activity.

If we adopt a simplified representation of the company's environment which would include customers, suppliers, investors, competitors and catalyst organisations, we can represent the main concerns of these protagonists.

One can easily notice that the above described relationships can be structured by contracts, but only in part (Bradach and Eccles, 1989; Medjad, 1996). In fact, trust stands behind formal agreements which are merely modelling a given activity.

Beyond the traditional customer-supplier relationships and the links between a firm and its different investors, informal or *tacit links* among competitors are also noticeable. In fact, they must cooperate to ensure the harmonious development of their industry. This is particularly true for companies working in emerging markets.

As for the role of institutions, among which we have identified the action of catalyst-organisations[9], their action consists in promoting and securing the different transactions.

Their involvement, as third parties, is actually crucial for the emergence and the existence of trust in the sphere of economics (Zucker, 1986; Shapiro, 1987; Sabel, 1993; Orléan, 1994; Coriat and Guennif, 1996).

These observations lead us to the following model (Figure 11.2):

Although there is scant literature on the role of trust in the creation of innovative companies (Sanner, 1996; Sapienza and Korsgaard, 1996), the concept of trust – or distrust – is rather familiar to researchers and practitioners in the field of strategic management. The 'elusive notion of trust' (Gambetta, 1988) even seems to be one of the most promising fields of research in the future.

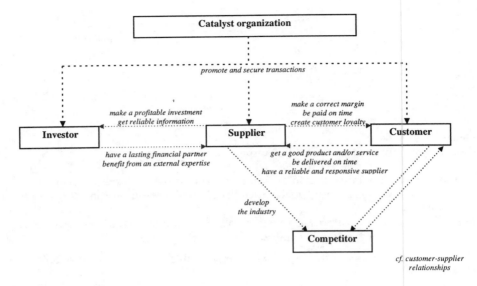

Figure 11.2
Trust in the environment of the firm: coexistence of open and focused relationships

The 'modern' principles and practices of management seem, actually, to be 'distrust-based' or, at least, based on power relationships that show a zero-sum view and a pessimistic view of economic agents. For example:

- *The transaction costs economics* (Williamson, 1975), which explicitly postulates the risk of an economic agent's opportunistic behaviour. The consequence of this theory has been the internalisation of a growing number of strategic transactions, the internal hierarchy being considered to be more efficient, or at least less risky, than market regulation,
- *The agency theory* (Fama and Jensen, 1983), which also postulates the opportunistic behaviour of managers as agents towards the shareholders which, in this case, aims at managing this risk through incentives and control mechanisms,
- *The population ecology of organisations theory* (Hannan and Freeman, 1977), ultimate symbol of the zero-sum game, where several organisations compete in an environment where the resources are durably limited. Cooperation is useless and trust is irrelevant in such a context,
- *The resource dependence theory* (Pfeffer and Salancik, 1978), where the organisation is considered as an open system which has to control its environment. Again, success does not have anything to do with trust[10],
- *The practice of real time management control*, in comparison with more open systems, where control is realised *ex post*,
- *The compulsive use of written contracts in western countries*[11], where all the probable futures are specified in order to prevent any hazard and any opportunistic behaviour of a partner,
- The relationships of pseudo-cooperation that can sometimes be observed between large companies and smaller ones, especially with new innovative firms, where the strict battle of wills often prevails, instead of the common creation of value.

In contrast, some ways of thinking or management modes are explicitly based on trust and, since the beginning of the nineties, meet a growing audience among western managers and scholars. This could be the beginning of a real paradigm shift. For example this might involve:

- *Some new theories of strategic alliances* and interorganisational cooperation (Ring and Van de Ven, 1992; 1994; Thiétart and Vandangeon, 1992) where mutual trust, based on experience or on shared norms and values, is analysed as a kind of implicit control,
- *The modes of management based on the principle of subsidiarity*, where an organisation does not reproduce a function which is already correctly carried out elsewhere,
- *The trend towards the abolition of intermediate hierarchical levels* inside an organisation, that shows the reduction in *ex-ante* controls (i.e. the everyday distrust), which might be replaced by *ex-post* controls, which do not require much unproductive manpower,
- *The 'network-organisations'*, where coordination is more informal than contractual or hierarchical.

According to Granovetter (1985), Zucker (1986), Peyrefitte (1995), Yoneyama (1995) and Medjad (1996), trust cannot be studied independently of its social and cultural environment. Its existence, its conditions of emergence and its different modes can hardly be dissociated from the period, the culture and the population in which it occurs.

The difficulty of analysing the concept of trust partly explains why many researchers have preferred to ignore it, rather than include it in their theories. Due to the conceptual hurdles encountered when analysing this metaconcept, management science has only considered trust as an 'invisible institution' (Arrow, 1974) or, sometimes, as a useless concept that 'only muddies the clear waters of calculativeness' (Williamson, 1993).[12]

RESEARCH PROCESS

One of the main difficulties in the study of trust consists in the ontological status of trust. It can in fact, be considered as:

- a *relationship*, which can be observed *ex post* or *ex ante*,
- a *cause*, leading to specific results,
- the *consequence* of a given interaction,
- a *construct*, involving an abstraction resulting from a model,
- a *process*, which can be broken down into stages.

Moreover, we have found that trust could hardly be separated from its immediate environment. Fundamentally, we should better strive to *understand* the concept of trust, rather than analysing, measuring or comparing it.

Studied population and selected sample

New innovative firms are seldom created with a complete team of managers and employees. In fact, a new company can often be reduced to the personality of its founder(s). This is the reason why we chose to focus on the interorganisational level, without asking the opinion of the employees who do not belong to the founding team. Among the 12 interviewed persons, we consequently gathered their opinion of the managers of 4 new companies.

The trust relationship being dual, we also intended to interview representatives from within the new firm's environment, including banks, customers, suppliers, catalyst-organisations and possible competitors.

Although we could not study the case of trust relationship between 2 competitors, we interviewed representatives of 4 other relevant groups, comprising: customers, suppliers, investors and catalyst organisations.

Our approach can be summarised in Figure 11.3 as follows[13]:

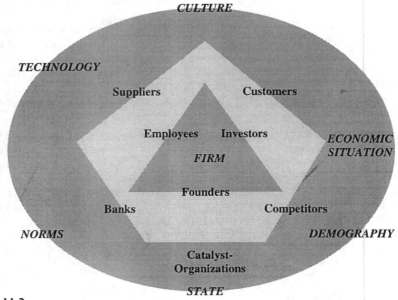

Figure 11.3
The firm, its environmental actors and the main factors influencing its development

Main steps of the research

Following a review of the literature, which provided an initial theoretical background, we completed a first series of 4 semi-structured interviews with managers of new innovative companies and members of their corresponding environments.

These interviews enabled us to gather various pieces of information and points of view with which to design the first questionnaire. We added questions on the possible factors that might vary between firms and, for the entrepreneurs, questions on self-confidence.

The synthesis of these 12 interviews – which lasted from 2 to 3 hours each – focuses on 9 specific topics: the definition of trust, the conditions for its existence, its limits, its value, its possible variation, its connection with risk, the conditions for its termination, its transitivity (the role and the opinions of other people) and its perceived reflexivity (why does a given person trust me?).

RESULTS AND DISCUSSION

The omnipresence of trust in the process of creation of an innovative company

The first question is '*Do you think that trust is important for the birth and success of new innovative firms?*'. All the answers are positive and one even added that trust was '*absolutely essential, in an etymological sense*'.

Yet this first result, which is not surprising, suggests another argument for those economists, sociologists or specialists in marketing and management science who consider that the notion of trust cannot be reduced to other concepts like interest (Lewis and Weigert, 1985; Servet, 1994; Hosmer, 1995).

Beyond this striking unanimity, it is however mentioned that:

- Trust is *'an important condition but is neither necessary nor sufficient: this does not prevent people whom everybody mistrusted from being successful'*,
- Sometimes, distrust is of greater importance: the problem will be more to dispel an *a priori* distrust than to create trust,
- Even if, in international trade, we trust a foreign partner, we need, however, to materialise the agreement through a contract so that many points, which cannot remain implicit, are specified because the commercial partner does not necessarily share the same values and, hence, the same perception of trust.

Rather different perceptions among the actors

On the definition of trust itself and its usefulness

After this preliminary question about the importance of trust, the best way to raise a definition of this concept was to ask directly the interviewed people to give us their own definition of trust and, thereafter, what was, according to them, the usefulness of trust, or its value, expressed in terms of difference of treatment.

The answers reveal the following:

- The notion of *relationship* explicitly appears on several occasions,
- The *non-strictly rational dimension* of this relationship is also asserted: in some definitions, trust is even opposed to a rational management of the relationship, emotions being in this case more important than cognitive aspects,
- If trust can be observed at a given moment, it is mainly for what it is going to allow in the future: half of the interviewed persons gave a similar definition: trust is therefore *a bet on the future*, according to a present observation,
- The definitions frequently mention *human qualities* like honesty, loyalty or morality.

We also asked the question of *the value of trust*. Through this interrogation, we wished to better understand the possible difference of treatment that could be given to entrepreneurs.

This differential reasoning helped us to seize the value of trust through concrete actions, beyond a sole empathy which, *per se*, would not be of any use for an entrepreneur.

We first noticed that the answers were clear-cut: without trust, almost nothing is possible. We particularly observed that:

- Trust helps in overcoming some apprehensions that would inhibit, or at least slow down, the decision-making process and the action: the first value of trust is *time*, an eminently rare commodity, as everybody knows. This materialises through a greater speed of processing for numerous types of requests including commercial propositions, credit, investment or subsidy, request for consulting, or technical assistance,
- The second important dimension is the *implication* of the person who trusts somebody. Besides a faster processing of applications, this implication materialises through *a clearly assumed risk-taking*: allocation of financial resources[14], and access to a professional network.

The main modes of interorganisational cooperation can, of course, be derived from these intangible assets such as time and risk-taking but, above all, cooperation has few opportunities to be effective without trust. These results are consistent with Mayer, Davis and Schoorman's (1995) model of trust and risk-taking in a given relationship.

About the conditions required to place trust in someone
Even if there are significant differences from one answer to another, trust appears to be placed in someone according to a subtle mix of objective and subjective criteria. This mix seems to result from one's personality, experience and level of responsibility.

Among the objective conditions, the intrinsic qualities of a new product or service are sometimes quoted. But subjective criteria prevail since the evaluation is made through '*a series of small signs*'.

We can postulate that an inference is often made between the viability of the firm and the characteristics of its founder, in whom we shall, all things considered, decide whether or not to place our trust. As a result:

- positive or negative, formal or informal information relating to the entrepreneur himself[15] will be a determinant of such a decision,
- information relating to the firm itself will often be perceived as a reflection of the professional and human qualities of its founder, and will also help decide whether or not to place one's trust.

The notions which often occurred in the different statements are:

- *prior elements* such as, on one side, education and *experience* and, on the other side, the absence of negative signs in this field. These elements allow us to infer the competence of the potential partner and his general intelligence,
- *immediate perceptions* like transparency, honesty, implication, tenacity or even some kind of seduction of the entrusted person. All these different elements create a certain kind of real or perceived *proximity*,
- *formal or informal guarantees*: written contract or tacit agreement. These future 'guarantees' are often associated with notions like reciprocity or capacity for giving.

Among these conditions, we also found the notion of self-confidence, without which a given evaluation would not be followed with action.

The main contribution of this part of the paper is to illustrate what also appears, although in a dispersed manner, in the managerial literature implying that the links of trust are based on past, present and future elements.

These elements can be gathered within the following triad of:

- *experience* (relative to the past),
- *perceived proximity* (present elements),
- *guarantees* (relative to the future).

Although trust has not yet been analysed in the light of such a triad, several authors have already mentioned these antecedents, including experience (Deutsch, 1960; Kee and Knox, 1970; Granovetter, 1985; Lewis and Weigert, 1985), perceived proximity (Lorenz, 1988; Sitkin and Roth, 1993; Thomas and Ravlin, 1995) and guarantees (Zucker, 1986; Shapiro, 1987; Medjad, 1996).

On the limits of trust and the conditions for its breach

The limits of trust If trust is a relationship based on several factors – experience, proximity and guarantees – it can, however, have some limits.

Except in 2 statements, which postulated that trust could be unlimited, a majority of the interviewed persons declared that there were obviously limits and that '*trust does not kill reason*'. We even noticed that the 2 persons who asserted that trust could be unlimited behaved accordingly, but only as long as trust existed. The sole mention of this restriction by the people concerned shows that there are, in fact, limits to trust even if they can only be observed *ex post*.

Concerning the real nature of these limits, we saw that the different answers reinforced the evoked conditions to place one's trust in somebody else. In order to study thoroughly this point, we consequently examined the cases of breach of trust.

Breach of trust: 'treason' as a frequent trigger   Following the conditions evoked to place one's trust, breaching factors are also numerous. Our question was '*Under which circumstances did you stop having trust in an entrepreneur or a project?*'[16].

The different answers clearly showed that the end of a trusting relationship occurs when the conditions of its attribution are no longer respected, as soon as one or more axes of our triad 'experience – proximity – guarantees' become questionable for the person who, until this moment, placed his trust in someone else.

Frequent allusions to, on the one hand, the notion of '*tacit agreement*' and, on the other hand, the notion of '*treason*' should be pointed out in that a trusting relationship frequently ends in cases of disloyal behaviour.

The relationship between trust and risk   Since many authors mention that trust allows risk-taking[17], we explicitly asked if the degree of trust could vary according to the level of risk incurred in a given situation.

The answers showed that, in many cases, trust is not adjusted to the level of risk but rather the opposite. According to the 'degree of trust' placed in a given person, one is inclined to take a certain level of risk. In other words, it is not the real or perceived level of risk which alters the trust placed in a given person, even if the frequency and the intensity of future validations often become more important.

Although the perceptions of some actors could vary significantly, it is, however, possible to find some regularities in the definition of trust itself, as well as in the processes required for its emergence, and the evolution and possible breach of a trusting relationship.

The crucial role of catalyst-organisations for the construction of trust

Trust is partly transitive

In order to study whether trust is totally or only partly transitive[18] and what could be the role of catalyst-organisations in the emergence of trust between 2 parties, our question was '*Are you sensitive to the opinion of other people in order to place your trust in a potential partner? If yes, until which point?*'.

The answers showed that, except in 2 cases[19], transitivity was not total, the final decision to place or not place trust in someone else or, on the contrary, to end trust in someone, remained eminently personal. On the other hand, there were, in several cases, facilitators and precursors of trust which played a significant role in the decision process.

The catalyst-organisations, key-actors for the birth of innovative companies

The partial transitivity of trust represents an interesting contribution because it shows potential relays in which a whole network of influencers, whose opinion about one or several dimensions of an innovative project, may be determinant.

This is, for example, true for the case of Anvar – the French national innovation agency, whose in-depth project analyses frequently determine the decisions of banks or private investors on investment. This is also the case for Sofaris, the main French guarantee-fund for small companies. Its decision to grant its guarantee to a given firm is also important, beyond the sole mathematical reduction of risk resulting for a bank or an institutional investor becoming involved.

This observation is remarkable because the partial transitivity of trust appears, not only to be possible between people but also between firms and *institutions*.

Even if the opinions expressed by these institutions result from the work of 'real people', this result shows the symbolic usefulness of what we call catalyst-organisations, which have, beyond their concrete mission, a label-effect that contributes to the emergence of trust in the economic sphere in general and in the environment of innovation and business creation in particular.

One of the essential qualities of a catalyst is, paradoxically, stability since it is a triggering factor of a 'chemical reaction', although it remains unchanged at the end of this reaction. This characteristic leads us to propose that new metaphor for a type of organisation which, without being directly involved, is nevertheless necessary to facilitate a commercial transaction.

The catalyst-organisations also seem to play a role in the development of *self-confidence*, which is an important factor in the decision to create a company and to carry on efforts towards innovation, in spite of problems inherent to every new venture.

The real or symbolic support of this type of how organism can help an entrepreneur to better overcome the periods of doubt which punctuate the innovation process.

SUMMARY OF THE RESULTS OF THE RESEARCH

Trust develops itself in three directions: proximity, guarantees and experience

A review of the managerial literature helped us to reveal several constitutive elements of trust, which could be confirmed and augmented with help of our different interviews.

We gathered these elements around 3 main axes:

- *experience* (relative to the past),
- *perceived proximity* (present elements),
- *guarantees* (relative to the future).

These elements represent facilitators or – if they work in the opposite direction – inhibitors of trust. We can summarise these factors in Table 11.1:

Table 11.1: *Facilitators and inhibitors of trust*

Facilitators of trust	Inhibitors of trust
Significant elements prior to the transaction	*Negative past significant elements (experience)*
Positive past experience	Negative past experience
Lasting relationship (duration)	Recent relationship
Positive reputation	Negative reputation
Proximity and similarity of the parties	*Distance and dissimilarity of the parties*
Geographical proximity between the parties	Important distance between the parties
Imminence of the closing of the transaction	Remote transaction term
Cognitive similarity	Cognitive dissimilarity
Common culture and shared values	Lack of common culture and shared values
Low level of responsibility of both parties	High level of responsibility of at least one party

Symmetrical relationship	Asymmetrical relationship
Presence of guarantees and elements of validations	*Absence of guarantees and elements of validations*
Formal contract or guarantees between both parties	No formal contract or guarantees between both parties
Existence of formal validation procedures	Absence of formal validation procedures
Existence of a formal or symbolic mediator	No formal or symbolic mediator
Low level of specific risk / minor stake	High level of specific risk / major stake
High quality and quantity of information	Poor quality and quantity of information
Respect of confidentiality	Absence of confidentiality
Existence of frequent and regular validations	No frequent and regular validations
Easy revocability of the transaction	Uneasy revocability of the transaction

Graphically, if we represent each axis as a dichotomy, we obtain the following depicted in Figure 11.4:

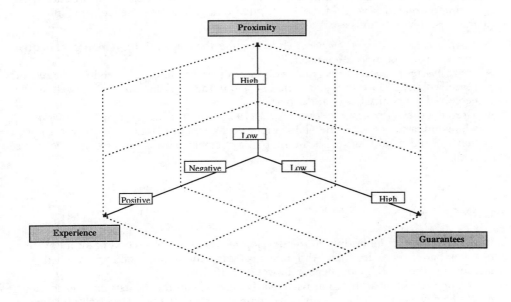

Figure 11.4
A framework of trust

Towards a definition of trust

In the light of the above dimensions, we can now suggest a general definition of trust, its antecedents, and its outcomes, which summarises the main elements encountered in the managerial literature and during our personal research:

Trust is *a positive anticipation associated with an assumed risk*. This relationship enables action and helps project oneself in the medium and long term: investment, cooperation – as opposed to instantaneous transactions – all the strategic moves that constitute an undertaken risk taking and which, due to their adoption, enable learning, organisational change and, by extension, progress and performance. Referring systematically to a third party, this relationship, which can be facilitated by catalyst organisations, is based on *experience* (involving prior elements) real or perceived *proximity* (including present elements) and future, (or *guarantees* related to the partner). This relationship lasts as long as the partners behave loyally, and ends in cases of perceived treason.

Although the issues of innovation and business creation are particularly subject to the question of trust, we considered that a definition, strictly limited to this field, would be irrelevant. We actually would argue that if the problem is exacerbated, in the case of an innovative new company, there are no specific rules which could not be transposed to the whole economic sphere[20].

Proposition of a trust-building mechanism

In order to 'produce trust' or at least to maximise its possibilities of emergence, we suggest emphasis on both subjective and objective levels.

We consequently have identified several types of actions designed to produce trust by acting on one or several of the 3 identified dimensions: experience, proximity and guarantees. These actions are:

- *Facilitating the information flow*, while keeping the critical information confidential (e.g. the heart of innovation itself)[21],
- *Increasing the frequency of interactions* between partners in order to reduce the real or perceived risk. More frequent contacts favour the information flow as well as the perceived proximity between the parties,
- *To secure each interaction by the use of a third party* as a real or symbolic guarantor, who is impartial and reliable. This might be a guarantee fund, or catalyst organisation[22] – and would formalise the mutual agreement with a contract or a symbolic action.

CONCLUSION

In the chain of factors leading to the development and to the success of an innovative new company, there are many different necessary conditions, sometimes called 'key factors of success'. None of them is, however, sufficient.

Yet, one condition seems to be particularly determinant for the harmonious development of a new venture which is the entrepreneur himself, the new company often being identified with its founder. This entrepreneur will, in fact, have to gather all the required energies to develop his company, even if, as we have seen, he can get support from associates and/or catalyst-organisations, whose influence will often be decisive.

Provided that the entrepreneur has enough self-confidence and is able to create a climate of trust for his project as well as for himself, notwithstanding some imponderables, the firm is likely to be successful.

While the independence of entrepreneurs increased with the help of new technologies of access, processing and transmission of data, their interdependence within an increasingly complex environment imposes real trust-building strategies, even if it is necessary to avoid to transform this requirement into a set of rhetorical tricks having little to do with the intrinsic qualities of the company and its manager.

However, we have noticed that, after a while, the true personalities are always revealed and that some kind of natural decantation takes place, the entrepreneurs showing, or not showing, their adaptation skills.

We always found that our triad experience-proximity-guarantees were implicitly completed with specific levels of risk and interest, and regulated by institutions like catalyst-organisations.

In any case, a relationship of trust is not neutral and contains *self-fulfilling prophecy components*. If you distrust a potential partner, you will have every chance to notice, in return, a behaviour pattern which will, *ex-ante*, justify your distrust. If, on the opposite, you trust someone (within reasonable limits) the self-fulfilling prophecy phenomenon will also work but, this time, in a positive direction.

NOTES

1. I would like to thank Karim Medjad for his helpful comments on an earlier draft of this essay.
2. The concept of trust stems from the Latin root *confidentia*. It can also be connected with the Roman *fides*.
3. cf. Van Wijk, G. (1996), *Trust and Structure*, ADSE Conference, Aix-en-Provence, April 1996.
4. cf. Thomas, D.C. and Ravlin, E.C. (1995), *Responses of employees to cultural adaptation by a foreign manager*, Journal of Applied Psychology, Vol. 80:1, pp. 133–146.
5. cf. Coriat, B. and Guennif, S. (1996), *Incertitude, confiance et institutions*, ADSE conference, Aix-en-Provence, France, April 1996.
6. cf. Ring, P.S. and Van de Ven, A.H. (1992), *Structuring cooperative relationships between organisations*, Strategic Management Journal, Vol. 13, pp. 483–498.
7. cf. McAllister, D.J. (1995), *Affect- and Cognition-based Trust as foundations for interpersonal cooperation in organisations*, Academy of Management Journal, 38:1, pp. 24–59.
8. Bidault, F. and Jarillo, J.C. (1995), *La confiance dans les transactions économiques*, in Bidault, F., Gomez, P.Y. and Marion, G. (1995), *Confiance, Entreprise et Société*, Paris: Editions Eska, pp. 109–123. Cf. also Sitkin, S.B. and Roth, N.L. (1993), *Explaining the limited effectiveness of the legalistic 'remedies' for trust / distrust*, Organisation Science, Vol. 4:3, August 1993, pp. 367–392.
9. A catalyst is defined as *'something that makes a chemical reaction happen more quickly without itself being changed, or, more generally, an event or person that causes great change'*. Cf. Zucker's (1986) analysis on the role of institutions in the development of trust.
10. This analysis should certainly be tempered: trust and control are not necessarily antagonistic. Cf. Bradach and Eccles (1989), or Sapienza and Korsgaard (1996).
11. *'Between American people, who prepare a contract as thick as a telephone directory and the Japanese, who consider that the best warranty is a trusting relationship, the gap is considerable'* (E. Yoneyama, 1995). Cf. also Medjad, K. (1996), *Contrat et confiance: une application à la Russie contemporaine*, ADSE conference.
12. Williamson, O.E. (1993), *Calculativeness, Trust and Economic Organisation*, Journal of Law and Economics, vol. XXXVI, April, pp. 453–486. Cf. *'I submit that calculativeness is determinative throughout and that invoking trust merely muddies the (clear) waters of calculativeness'* (p. 471).
13. This diagram is partly derived from Joffre, P. and Koenig, G. (1992), *Gestion stratégique. L'entreprise, ses adversaires-partenaires et leur univers*, Paris: Litec (p. 138).
14. cf. the notion of *credit* mentioned by 2 respondents: its etymology directly leads us to the Roman '*fides*'.
15. or relative to the team of entrepreneurs, if it exists.
16. For the entrepreneurs, the question was *'Under which circumstances did you stop trusting a partner of your company and/or of your innovation?'*.
17. cf. Ring and Van de Ven (1992, 1994), as well as Mayer, Davis and Schoorman (1995), or Coriat and Guennif (1996).
18. In other words, this leads us to study the following proposition: in which extent, if A trusts B and if B trusts C, then A trusts C?
19. The same persons who considered that trust was unlimited.

20. Since a new company has no past, trust can be expressed in terms of direct relationships between the entrepreneur himself and the different components of his environment.
21. The information which could be circulated faster is related to the norms, the prior experience of the team of entrepreneurs, the technology itself, the targeted markets and the competition: everything that helps making a decision with a full knowledge of the facts. The objective is to allow a fast and less expensive validation, which implies a diminution in transaction costs (cf. Williamson, 1975).
22. It should be pointed out that catalyst-organisations have an impact on the 3 dimensions of trust: experience, proximity and guarantees. The stake is thus to enshrine the parties into a network of economic partners, in order to provide more visibility to the mutual cooperation.

REFERENCES

Albert, P. et Mougenot, P. (1988) La création d'entreprises high tech, *Revue Française de Gestion*, Mars-avril-mai, pp. 106–18.
Arrow, K.J. (1974) *The Limits of Organisation*, Norton, New York.
Bidault, F. and Jarillo, J.C. (1995) La confiance dans les transactions économiques, in Bidault, F., Gomez, P.Y and Marion, G., *Confiance, Entreprise et Société*, Eska, Paris.
Bradach, J.L. and Eccles, R.G. (1989), Price, authority and trust: from ideal types to plural forms, *Annual Review of Sociology*, 15, pp. 97–118.
Coriat, B. and Guennif, S. (1996), Intérêt, confiance et institutions. Une revue de littérature, ADSE Conference, Aix-en-Provence, April.
Fama, E.F. and Jensen, M.L. (1983) Separation of Ownership and Control, *Journal of Law and Economics*, 26, pp. 301–25.
Gambetta, D. (ed.) (1988) *Trust: Making and breaking cooperative relations*, Blackwell, Oxford.
Granovetter, M.S. (1985) Economic action and social structure: the problem of embeddedness, *American Journal of Sociology*, 91, pp. 481–510.
Hannan, M.T. and Freeman, J.H. (1977) The population ecology of organizations, *American Journal of Sociology*, 82, pp. 929–64.
Hosmer, L.T. (1995) Trust: the connecting link between organisational theory and philosophical ethics, *Academy of Management Review*, 20, 2, pp. 379–403.
Kee, H.W. and Knox, R.E. (1970) Conceptual and methodological considerations in the study of trust and suspiscion, *Journal of Conflict Resolution*, 14, pp. 357–66.
Lewis, J.D. and Weigert, A. (1985) Trust as a social reality, *Social Forces*, 63, pp. 967–85.
Lorenz, E.H. (1988) Neither Friends nor Strangers: Informal Networks of Subcontracting in French Industry, in Gambetta, D. (ed.) *Trust: Making and Breaking Cooperative Relations*, pp. 194–210, Blackwell, Oxford.
Mayer, R.C., Davis, J.H. and Schoorman, F.D. (1995) An integrative model of organisational trust, *Academy of Management Review*, 20, 3, pp. 709–34.
McAllister, D.J. (1995) Affect- and cognition-based trust as foundations for interpersonal cooperation in organisations, *Academy of Management Journal*, 38, 1, pp. 24–59.
Medjad, K. (1996) Contrat et confiance: une application à la Russie contemporaine, ADSE Conference, Aix-en-Provence, April.
Orlean, A. (1994) Sur le rôle respectif de la confiance et de l'intérêt dans la constitution de l'ordre marchand, in *A qui se fier? Confiance, interaction et théorie des jeux*, La revue du M.A.U.S.S., N°4, second semestre 1994, pp. 17–36.
Peyrefitte, A. (1995) *La Société de Confiance – Essai sur les origines et la nature du développement*, Editions Odile Jacob, Paris.
Pfeffer, J. and Salancik, G. (1978) *The External Control of Organisation: A Resource Dependance Perspective*, Harper and Row, New York.
Ring, P.S. and van de Ven, A.H. (1992) Structuring cooperative relationships between organisations, *Strategic Management Journal*, 13, pp. 483–98.
Ring, P.S. and van de Ven, A.H. (1994) Developmental processes of cooperative inter-organisational relationships, *Academy of Management Review*, 19, 1, pp. 90–118.
Sabel, C.F. (1993) Studied trust: building new forms of cooperation in a volatile economy, *Human Relations*, 46, 9, pp. 1133–70.

Sanner, L. (1996) Trustbuilding between entrepreneurs in new ventures and external actors, ICSB World Conference, Sweden, 17–19 June.

Sapienza, H.J. and Korsgaard, M.A. (1996) Procedural justice in entrepreneur-investor relations, *Academy of Management Journal*, 39, 3, pp. 544–74.

Servet, J.M. (1994) Paroles données: le lien de confiance, in *A qui se fier? Confiance, interaction et théorie des jeux*, La revue du M.A.U.S.S., N°4.

Shapiro, S.P. (1987) The social control of impersonal trust, *American Journal of Sociology*, 93, 3, pp. 623–58.

Sitkin, S.B. and Roth, N.L. (1993) Explaining the limited effectiveness of the legalistic 'remedies' for trust / distrust, *Organisation Science*, 4, 3, August, pp. 367–92.

Thietart, R.A. and Vandangeon, I. (1992) Direction et contrôle des alliances stratégiques, in *Mélanges en l'honneur de Jean-Guy Mérigot*, pp. 577–590, Economica, Paris.

Thomas, D.C. and Ravlin, E.C. (1995) Responses of employees to cultural adaptation by a foreign manager, *Journal of Applied Psychology*, 80, 1, pp. 133–46.

Van Wijk, G. (1996) Trust and Structure, ADSE Conference, Aix-en-Provence, April.

Williamson, O.E. (1975) *Markets and Hierarchies, Analysis and Antitrust Implications*, Free Press, New York.

Williamson, O.E. (1993) Calculativeness, Trust and Economic Organisation, *Journal of Law and Economics*, 36, April, pp. 453–86.

Yoneyama, E. (1995) La relation de confiance dans les affaires au Japon, in Bidault, F., Gomez, P.Y. and Marion, G. (eds.) *Confiance, Entreprise et Société*, Editions Eska, Paris.

Zucker, L.G. (1986) Production of trust: institutional sources of economic structure, 1840–1920, in Staw, B.M. and Cummings, L.L. (eds.) *Research in Organisational Behavior*, 8, pp. 53–111, JAI Press, Greenwich, CT.

PART VI   Co-operation Between Small and Large Firms

CHAPTER 12

Dynamic Complementarities with Large Advanced Companies: The Impact of their Absence upon New Technology-Based Firms

MARGARIDA FONTES

INTRODUCTION

The establishment and growth of New Technology-Based Firms (NTBFs) have been associated with the presence of technology oriented organisations, which act as technological 'incubators' (Cooper, 1985) and with which NTBFs can establish 'dynamic complementarities' over time (Rothwell, 1983). These organisations include research institutions and large technology intensive companies. This paper is particularly concerned with the latter, which were found to have an important impact, both at the level of supply – providing entrepreneurs, technology and complementary assets – and at the level of demand – absorbing a substantial proportion of NTBFs products and services.

NTBF creation frequently results from the identification and exploitation of technological opportunities that are ignored or neglected by existing organisations. NTBF founders are, to some extent, 'internalising' into a new organisation the externalities generated into the process of creation and diffusion of knowledge in other larger firms (Fontes, 1997). In some fields, large advanced companies have consistently been a major source of the knowledge and skills used by NTBF founders to launch their firms. The phenomenon of large firms 'incubating' technical entrepreneurs is evident in the US (Rothwell, 1983; Cooper, 1985; Bahrami and Evans, 1995). It has also been identified in European studies (Kulicke and Krupp, 1987; Pottier, 1988; Granstrand and Sjolander, 1990) although it has been argued that conditions in Europe are less conducive to such 'incubation' (Roure and Keeley, 1989). Such a 'spin-off' perspective does not necessarily mean a direct transfer of a technology or a product. But at the very least, ideas, skills, and contacts are likely to be derived from the source organisation (Roberts, 1991). Indeed, NTBF founders tend to launch their firms in the same industry/market as the companies where they worked before, use a similar technology, and carry out a similar type of work (Cooper, 1985; Oakey and Cooper, 1991).

In some fields such as biotechnology (Kenney, 1986; Walsh, 1993) or laser technologies (Shearman and Burrell, 1988), knowledge exogenous to established firms, and largely developed by universities and other public bodies, appears to be more relevant. Similarly, in less advanced countries, where the industrial structure is weaker, the technological infrastructure from within public bodies is the principal source of new knowledge (OECD, 1992). In these cases, existing firms are less influential with respect to the generation of knowledge and skills. But whether or not they are a source of technological knowledge, large technology intensive firms can play a number of other roles extremely relevant to NTBFs. They may provide resources the new firm does not possess – such as capital or other 'complementary assets' (Teece, 1986) – and, because they have sophisticated needs, they may be sources of demand for its products.

Large firms can act as providers of start-up capital (usually in the form of minority shareholdings) by assisting ex-employees who create a firm to exploit an idea/technology developed in-house, or by supporting independent entrepreneurs. Their interest is often to maintain a close observation of new technologies being developed, with a view to eventually complementing their own efforts with these developments, or to gain access to completely new areas of technology (Pottier, 1988; Olleros and MacDonald, 1988; Roberts, 1990). The degree of an NTBF's autonomy varies, but many of these firms are eventually acquired by a larger partner (Granstrand and Sjolander, 1990; Roberts, 1990). Large sponsoring firms may also provide other forms of assistance, such as managerial support, or access to manufacturing facilities and distribution channels (Rothwell, 1983; Walsh, 1993).

The access to assets they do not possess may be critical to NTBFs, which usually have limitations with respect to resources and breadth of skills. Evidence points to a deficiency of business skills and lack of experience in non-technical areas, with particular relevance to marketing (Oakey, 1991; Roberts, 1991). Thus, relationships with large firms are often driven by the need to obtain missing skills or assets, with the NTBF trading its specialised competencies for them (Lawton Smith et al., 1991). One basic conclusion derived from the literature on large/small firm relationships is that the main advantage for the small firm lies in access to these resources, rather than from the acquisition of technological knowledge, which is often what the large firm seeks (Shan, 1990; Walsh, 1993). Nevertheless, because NTBFs tend to have narrow specialisations and thus need to obtain complementary knowledge, they gain from collaborative relationships, where the large partner covers a substantial part of research expenditures and may also benefit from a contact with the large firm's broader range of technology competencies (Lawton Smith et al., 1991).

However, relationships with large companies are bound to raise power problems, due to the enormous resource differences between partners. Small firms may face appropriation problems (Shan, 1990; Lawton Smith et al., 1991) and there is always the threat of acquisition, although it was observed that some large firms prefer alliances, because of the fear that key people might leave (Pisano, 1991). Thus, these relationships are often feared by the weaker partner, who only enters into them when forced to by lack of essential resources (Rothwell and Dodgson, 1991).

But possibly the most important function played by large advanced firms is acting as a source of demand. The existence of a market for the technically sophisticated products supplied and, especially, the willingness of existing companies to buy from firms that have not yet established a reputation, or in fields where the local industry is not recognised as having previous experience, is a determinant condition for the survival of NTBFs (Dahlman and Westphal, 1982; Walsh, 1993). According to Carlsson and Jacobsson (1991) users are as important as suppliers and if they 'fail to appear in large enough numbers the rate of diffusion of new technology can be retarded for the whole industry'. The limited demand for technology intensive products in less industrialised countries is a major obstacle to be faced by local firms trying to introduce new technologies (Dahlman and Westphal, 1982; Deniozos, 1994).

The above discussion is consistent with studies that stress the importance of a solid industrial structure for the formation and development of NTBFs (Monck et al., 1988; Kim and Dahlman, 1992). This means that, in less advanced countries, technological entrepreneurship is likely to be constrained by a number of 'locational disadvantages' (Perez and Soete, 1988). The limited presence of industrial participants in the process of technology generation restricts the emergence of technological opportunities and the number of people prepared to identify and exploit them, and constrains the extent of technological linkages that can be established at industrial level. The low technological awareness prevalent among the majority of local firms reduces the level of demand for technology intensive products. Thus the rate of NTBF creation is likely to be low and those created are often isolated in their efforts and confronted with an unfavourable environment, especially with respect to opportunities for market expansion (Fontes, 1997).

Therefore, it can be suggested that the characteristics of the industrial structure – and particularly the absence of large technology intensive companies – are a significant shortcoming in NTBF creation and development, excluding one important actor in the 'incubating' process and considerably limiting the establishment of the 'complementarities' described above. The objective of this paper is to analyse the impact of these conditions on NTBF creation and development. The case of Portugal – where there are very few large technology-based firms and where other technology-aware companies are less numerous than in advanced countries – is the setting chosen here to address this issue.

METHODOLOGY

The results presented below are based on an empirical study of NTBF creation and development in Portugal, carried out between 1991 and 1994. In-depth interviews were conducted with the founders of 28 young independent firms, created in the last 15 years and involved in the development and/or application of technologies new to Portugal. These firms were selected from the 123 firms that responded to a mailed questionnaire, addressed to a universe of Portuguese NTBFs (identified through an extensive search), which provided a first characterisation of this group of firms. Most firms studied operated in fields which can be broadly included under the 'information technologies' (IT) umbrella. This is related to the fact that the development of the market for IT provided the greatest opportunities for NTBF creation (Laranja, 1995).

Biotechnology firms were excluded from this first study, given the very low incidence of NTBF creation in this sector in Portugal, and considering that firms in these fields differ substantially from IT based ones (Oakey and Cooper, 1991). However, an ongoing project focusing on NTBF formation in biotechnology – involving case studies of the 11 firms it was possible to trace so far – provides some preliminary information, which may complement conclusions mainly derived from the IT area.

The paper addresses the role of existing companies in the formation and further evolution of NTBFs. It focuses on the impacts of the scarcity of established technology oriented companies upon the access to the 'inputs' critical to NTBF creation and development: namely technology, market demand and funds. Particular attention is given to the first 2 elements which were identified elsewhere as the 'weak links' for NTBFs in a less advanced country (Fontes and Coombs, 1996). The paper examines the impact of the absence of these important actors upon the development of local NTBFs, and attempts to identify the practices they developed to lessen such impact.

TECHNOLOGICAL INFLUENCE OF LARGE ESTABLISHED FIRMS

Identification of an Opportunity and Initial Technology Acquisition

An analysis of the founders' origin concluded that the majority (57.9%) came from existing companies – most of which could be described as large in the Portuguese context. Twelve firms were formed by people with exclusively industrial backgrounds. In another 9, formed by teams with different backgrounds, the main entrepreneurial role was still played by an individual originating from industry. Despite this predominance of industrial backgrounds, the role of Portuguese companies as technological 'incubators' was found to be very limited. In fact, the analysis of the way the technological opportunity was identified led to the conclusion that in only a very small number of cases was the previous industrial employer the most relevant technological influence on the new firm. However, some technology aware companies had afforded their employees the possibility to access knowledge generated in more advanced environments – in Portugal or abroad – and had provided them with the skills and knowledge necessary to recognise and exploit it. A few NTBF founders originated from local subsidiaries of foreign Multi-National Enterprises (MNEs), but with one exception, their role was similar to that of the technology aware Portuguese companies.

Industrial employers were found to have a greater contribution to the identification of market opportunities. The pertinence of this contribution becomes evident when one considers the case of firms launched on the basis of a potential technological opportunity, which spent a long time devising an application, or finding a market for it. The possibility that some company environments offered their employees opportunities to identify a market for the new product and/or to achieve a good insertion into the relevant trade networks, was an invaluable contribution for the early survival of NTBFs.

The influence of the ex-employer in the actual acquisition of the initial technology (as perceived by the founders), was also analysed. As Table 12.1 shows, *direct technology transfer* – i.e. the transfer of specific development results, or of an almost complete application – rarely took place between an industrial employer and the new firm, although this was a typical situation in the case of research organisations.

Table 12.1: *Initial technology acquisition – influence of ex-employer*

	Firm (or professional)	Research organisation	Other influence*
Direct technology transfer	2	5	3
Skills transfer determinant	10	2	–
Low influence	5	1	–
TOTAL	17	8	3

*Industry based founders who mentioned a relevant influence from an organisation other than the employer.

A situation where the *transfer of knowledge or skills* embodied in the individuals was perceived as *determinant* was the most frequent among industry based founders, who came from the same industry or had done similar work previously. But while a few founders mentioned the presence of knowledge that was directly applicable in the development of new products, more frequently they referred to the learning of technical skills, to the acquisition of market knowledge, and to a good integration into the sector in terms of access to suppliers, contacts with potential clients, and knowledge about distribution channels. The perception of a *low* technological influence of the ex-employer was typical of industry based founders, who had identified a potential need felt by the employer or by its clients, which they tried to meet through the development of a new technological application. This often entailed departing significantly from the areas where they had gained experience, requiring the acquisition of new knowledge and skills. For most of these founders no organisation could be identified as a *direct* source of the technology used or the skills applied, although in a small number of cases, depicted as *'other influence'*, the need was met through direct technology transfer from other organisations.

The question that remains to be asked is: considering the limited technological influence of established companies, how did NTBF founders identify the technological opportunities and where did they obtain the technology necessary to launch their firms? In particular, how did the industry founders who did not acknowledge a direct influence of their previous organisation achieve their launch? The research uncovered 2 alternative influences which were *local research organisations* and *foreign sources*.

Research organisations were more clearly technological 'incubators'. However, only a small proportion of founders originated from them, and very few organisations were at the origin of direct spin-offs. There were also cases of external entrepreneurs who gained access to technology developed in a local research centre, which they converted into a marketable application. Furthermore, some founders originating from industry established links with local research organisations which were instrumental in the development of the new product. A predominant influence of research organisations was the norm in the formation of biotechnology firms, where both the knowledge and the entrepreneurs originated from a research environment.

But some founders went beyond the local environment to identify opportunities, basing their firms, at least in part, on knowledge or technology obtained abroad. NTBF founders were part of an 'elite', who had access to a variety of sources of information and knowledge. This permitted them to be aware of technologies emerging in more advanced environments, and of their potential applications. For some of them, the information and/or knowledge thus accessed were the major influences on the final identification of a technological opportunity. Besides the channels provided by the ex-employer, other forms of access included knowledge and skills acquired through postgraduate research or work abroad, direct contact with foreign organisations, exposure to new knowledge through less tangible means: publications, international meetings and personal networks. In a few instances, the result of this process was the purchase of advanced technology or its licensing. But, in most cases, the materialisation of the opportunity identified entailed a greater involvement of the founders in the actual *development* of the technology. This process, which comprehended an important element of 'self-learning' and of trial and error experiment, involved combining public knowledge and 'spill-overs' from foreign research with existing skills and training in new ones, sometimes supported by informal links with local research.

The creation of firms strongly based on technological influences originating outside the country can be regarded as a way of overcoming the limitations of the country's national system of innovation. Some of these firms have acted as pioneers, contributing to introduce the new technologies in the Portuguese context. In gaining competencies in these 'absent' technologies, and in applying them to the development of products or services (often oriented to the needs of the local market), as opposed to simply purchasing and using them, NTBFs have acted as agents of their endogenisation at country level.

In performing this role, more than being influenced by established firms, NTBFs had an important technological influence upon them. In fact, as will be discussed below, NTBFs have had a relevant role as diffusers of the new technologies and their applications among a set of more aware users, mostly composed of large firms, which were prepared to recognise the relevance of the new solutions, but had no capacity to devise them.

External Relationships and NTBFs' Continued Technology Acquisition

Portuguese NTBFs, in common with firms elsewhere, have scarce internal resources and relatively narrow competencies (Fontes and Coombs, 1995). Therefore, they are likely to benefit from the possibility of supplementing their in-house technology acquisition efforts with inputs from external sources.

The research found that the role of large established firms was equally limited concerning NTBF knowledge acquisition over time. Opportunities to establish technological relationships with local industrial organisations (other than these of the user-supplier type) were constrained by the relatively small number of technology advanced companies. Some firms were truly 'lonely' and complained about the lack of local interlocutors and missed the presence of a technologically dynamic industrial community. Informal know-how trading (von Hippel, 1987) among R&D workers from different firms was mentioned in some fields, being encouraged by geographical proximity. However, formal relationships were rare. The growing number of new technology-based firms, as well as the activities of a few large firms and/or their technology-based subsidiaries, were contributing to an increase in opportunities for more extensive collaboration. However, the cases of collaboration identified were predominantly between small technology intensive companies.

Portuguese NTBFs also benefited less than similar firms elsewhere from the technological linkages that can be established in the context of close user-producer relationships (Lundvall, 1988). In some fields, NTBFs and a few other technology intensive firms functioned as suppliers of advanced (often customised) inputs to each other, generating a two-way movement of technological know-how, but this was not the prevailing situation. NTBFs used extensively foreign suppliers for sophisticated inputs that they could not find locally (sometimes through local agents, but often resorting to direct imports). With respect to customers, technological co-operation was seldom mentioned. Large clients were a source of ideas/problems to be solved and sometimes of customer requirements, but rarely of technical knowledge. Therefore, the relationship established frequently involved the one-way transfer of technology from the NTBF to its clients, with the technology intensive supplier effectively acting as disseminator of advanced technology amongst them.

In the case of biotechnology, ongoing research uncovered the limited interest in biotechnology processes and products revealed by most established firms from sectors potentially affected by this technology including food processing, pharmaceuticals and pulp and paper. Most firms in the first 2 sectors are small and have a moderate to low technological orientation, often operating under licence. Several were recently acquired by foreign companies, which reduced their margin for manoeuvre. The pulp and paste sector is dominated by a few larger and more technology advanced companies, that use up-to-date processes. But these processes are stabilised and the companies are reluctant to introduce new technologies that could interfere with their performance. Only a couple of large companies have some activity in the biotechnology field and a few others participate in, or at least finance, projects being carried out at the university. But these activities tend to be regarded as 'marginal' in relation to the firms' main business. This suggests that Portuguese biotechnology firms may have few chances of establishing the type of technological linkages that permitted similar firms elsewhere to access indispensable complementary assets (Pisano, 1991; Walsh, 1993).

Therefore, in accessing technological knowledge over time, NTBFs remained strongly dependent on the same sources they had used at their launch, which were research organisations and foreign sources. To these can be added linkages with other NTBFs, when such other firms existed.

In an environment where industrial research is scarce, relationships with universities and other research organisations assume a great pertinence for technology intensive firms. Some NTBFs had developed a close, sometimes 'symbiotic', relationship with a research organisation. This was the case of spin-offs from research or of firms created through technology transfer. But not all firms had such a positive attitude regarding research organisations. Some considered them useless, usually because the firm operated in fields where local research was absent, or was too 'academic' to be of use to it (but some still kept occasional links for 'scanning' purposes). Others, recognising the potential advantages of relationships, used them in an 'opportunistic' way and tended to adopt a careful posture with respect to collaboration, given the differences in objectives and behaviour. In the case of foreign sources, market relationships with sophisticated suppliers prevailed, but some firms had engaged in collaborative relationships, frequently as a result of commercial contacts or of participation in European research programmes. Several firms expressed the desire to establish partnerships of a technological nature with foreign companies. However, most NTBFs looked for equal partners such as other small firms, since they feared the power of large companies. In looking for these partners NTBFs searched for technological know-how and *also* industrial experience in general, something they lacked because of their youth and the limited industrial tradition in their fields in Portugal.

The alternative ways used by Portuguese NTBFs to access complementary knowledge and technology have been discussed in more detail elsewhere (Fontes and Coombs, 1995). It was concluded that, with time, most NTBFs were able to surmount the difficulties associated with technology access. However, as we will see below, the problems with respect to market demand were more serious and NTBFs faced greater difficulties in this area.

MARKET DEMAND

Large Firms as Clients

The vast majority of NTBFs were oriented to non-consumer markets and therefore their clients were other firms, or the public sector. Large national firms, and to a lesser extent MNE subsidiaries, in the service sector (particularly telecommunications, financial services and computing), and in a few industrial sectors, were identified as the principal clients of NTBFs (Table 12.2). According to Laranja (1995), the increase of market opportunities in the electronics and information technology sectors was associated with a process of technological upgrading and transformation undertaken by large local users. This led to the development of niche opportunities based on the needs of the domestic market, which involved both the development of products customised to local requirements and the adaptation of products and systems developed elsewhere.

Large technology aware companies played an important part in the early activities of NTBFs. The majority of the firms interviewed could trace the presence of one (or a few) clients that sustained their early steps, an association that occasionally progressed to a close user-supplier relationship. Some of these clients could be described as 'lead-users' since they absorbed a relevant part of firms' production, acted as a test bed for the product, contributed with ideas and suggestions, provided knowledge of custom requirements and, more rarely, technical inputs. This role was sometimes played by the ex-employer. Others could be simply described as 'early adopters' as they were willing to try the products and prepared to buy them from a firm without a past record, but in this case the relationship was less close. Finally, a few firms, because of the type of product being commercialised or the market targeted, almost immediately started selling for a wider market, although sometimes the initial product had been developed for a particular customer.

Table 12.2: *NTBFs' early and current markets by type of market*

	Early markets			Current markets		
	Large firms	SMEs	Other	Large firms	SMEs	Other
Total NTBFs	18	7[1]	11	22	8[2]	13
Type of market* *Manufacturing* *Services*[3] *Public*[4] *Military*	*13* *9* – –	*6* *1* – –	– – *11* –	*14* *13* – –	*6* *3* – –	– – *11* *3*

*Some firms operated in more than one type of market, so the sum is higher than the total of NTBFs
[1] Three of these firms also have large or public clients; [2] All firms also have large or public clients; [3] Include distributors; [4] Include public utilities (e.g. telecommunications; electricity).

But while some firms were able to procure clients with a steady demand, others faced serious problems with regard to market acceptance of the technologies they were introducing. Some of the latter were ultimately able to open up a market after long 'desert-crossings', but others were forced to change the contents of their activities. These adjustments included cases where firms had to move to another activity altogether, and cases where they persisted in the chosen business but were obliged to alter their products by 'downgrading' them to conform to the real needs of their potential clients, or to undertake other (usually less sophisticated) activities in parallel.

On the other hand, some initially successful firms were later confronted with the limitations of the market, when required to replace the early lead-users, or when further growth was envisaged. In fact, if the early adoption of the firms' products by a responsive customer eased the initial product launch, it was no guarantee of a wider market. The experience of older NTBFs has shown that the move beyond the early 'lead-user' can be difficult. Some younger firms remain very dependent on one large client, and it remains to be seen whether they will be able to undertake such a move (a couple of recent failures substantiates these fears). NTBFs are often described as experiencing difficulties when seeking to move beyond a few customers, because they lack the required marketing skills (Oakey, 1991). This is also the case with Portuguese NTBFs which share the weaknesses of their counterparts elsewhere which include resource constraints, absence of a reputation, and limited marketing skills. But, their task is likely to be even more complex, due to the gap between a few advanced clients and a multitude of less sophisticated and less trusting ones, which are much harder to gain as clients of technology intensive products.

One of the principal problems of Portuguese NTBFs is that the number of organisations prepared to assume the role of 'lead-user', or at least act as 'early adopter' of any new applications being introduced, is relatively small. Beyond an 'elite' of sophisticated users, NTBFs have to confront the low technological level, and the conservative behaviour, of 'average' companies, which are particularly strong in some sectors. The serious problems experienced by NTBFs which have attempted to direct the new technologies towards the upgrading of firms in traditional sectors show that, although national NTBFs may be better prepared than the foreign suppliers to adjust their technologies to the customers' needs (given their flexibility and their proximity) it is unlikely that NTBFs are easily accepted by these firms as suppliers. Unless there are some major changes in the behaviour of these potential customers, the client for the

Portuguese NTBF remains the large technology aware customer or the small technology intensive company. These are equally the principal clients of NTBFs elsewhere, but because in Portugal such firms are less numerous, the difficulties experienced by local NTBFs in gaining other customers can affect their potential for survival and growth.

Close Client-Supplier Relationships with Sophisticated Users

NTBFs that were able to target niches composed of more sophisticated users, experienced less early problems and had better conditions in which to expand their activities. A substantial proportion of these clients were not technology intensive, but revealed some interest in novel products incorporating advanced technologies, and were willing to buy them from local suppliers. Several of these advanced clients would have been able to resort to foreign suppliers, and satisfy their needs on the basis of imported technology. The fact that they *trusted* local firms instead favoured the building-up of national competencies in the development and use of the technologies involved. In return, they enjoyed the benefits of a close link with a local company, some of which are in the process of developing complex technologies (Lundvall, 1988). Advantages include, product customisation to customer needs instead of the use of a standard product; the advantage of being the principal client compared to being only marginal with regard to a large foreign supplier. Other advantages include proximity in terms of both development and after sales support and the guarantee of continuity of supply.

The establishment of these close relationships favoured a learning process. As a result, NTBFs could continue to expand their technological competencies and adjust them to the needs they identified in their markets. However, such adjustments were beneficial for the NTBFs, only when they targeted niches composed of demanding clients which allowed them to develop truly innovative products, or when the clients' requirements evolved over time, making it possible for them to assimilate increasingly complex products and services. When these processes took place, a sort of 'virtuous circle' was induced. If not, these adjustments could be dangerous for a firm's future prospects, particularly if foreign expansion was envisaged. In fact, the 'specific needs' of Portuguese clients could be relatively unsophisticated, and thus not serve as basis for international expansion.

Considering that aware clients were generally not technology-based firms or, at least, did not have technological competencies in the fields where they used the NTBF as supplier, it is possible to conclude that NTBFs contributed to the dissemination of new technologies amongst this set of users. The extent of this dissemination effort resulted in the impact of NTBFs upon 3 levels: (a) in bringing these technologies to the market they have first of all raised the awareness of them and their applications among potential users and have also made them readily available to a set of 'early adopters'; (b) in establishing a good user-supplier relationship with a set of informed users they have begun to diffuse advanced technology amongst these customers, by providing access to technology that they might not otherwise use, or by guaranteeing local supply of a technology they would otherwise import and by devising applications that satisfied their needs more closely; (c) the few firms that were also able to take the technology beyond this sophisticated group reached users that might not have been aware of the technology's potential if the firm had not made it available to them (e.g. adapting it to the customers' requirements, thus having a particularly important role as diffusers).

If we address this issue from the standpoint of NTBFs, it is also possible to reach some conclusions with respect to the impact of the good clients upon NTBF development. Clients with a steady demand may permit the development of the first product, guarantee the early survival of an NTBF, and provide an incentive for these firms to continue upgrading and expanding their technology base and passing the results to their clients. However, dependence on one or a few clients renders firms vulnerable thus threatening the continuity of their technology acquisition and diffusion efforts. So, the ability/possibility to extend their activities to a wider range of good clients (either by expanding the client base for the same product or its modifications, or developing new products by targeting other new niches) confers on NTBFs a greater stability. They are then better prepared to retain and expand their in-house technological competencies to provide benefits for their customers. They are also more reliable as suppliers, ensuring the continuity of their products and services.

The Role of Large Firms in NTBFs' Market Internationalisation

Given the limited demand for technology intensive products, foreign market expansion may be necessary for NTBFs to survive and grow as technology intensive companies. Internationalisation is always a difficult process for small firms (Coviello and Munro, 1995) and in the case of Portugal, NTBFs are still hindered by the image of low-technology producer, generally attributed to the country in international trade. Some firms were confined to the national market by the nature of their activities. The export prospects of the remaining NTBFs were, to some extent, still influenced by the conditions of the domestic market. Thus, some previous success in the Portuguese market was extremely relevant for firms engaging in foreign expansion because it provided resources, industrial experience, reputation and contacts. The nature of local clients' demand was also an important factor, since it influenced the characteristics of the products, and hence their potential success in demanding markets.

The ability to establish a network of supportive relationships that compensate for gaps in marketing knowledge and scarcity of resources was found to be critical for the success of NTBFs when attempting internationalisation. Among these, relationships with large companies have emerged as critical. Two situations were identified. In some cases, large clients supported the internationalisation efforts of their small NTBF suppliers, either through a partnership or by providing access to their commercial channels. In other cases, large Portuguese companies selling in foreign markets, or Portuguese subsidiaries of foreign MNEs, acted as NTBFs' intermediaries to foreign clients (Fontes and Coombs, 1997).

Biotechnology is an extreme example of demand constraints. The low rate of firm creation was invariably justified by entrepreneurs and researchers in the field by the absence of market opportunities. The local market did not seem to provide enough demand to motivate would-be-entrepreneurs, and had not permitted the development of the firms created in the 1980s. Therefore, early internationalisation seemed to be a requirement and some of the firms created in the 1990s were careful in trying to guarantee orders from *foreign* clients before launching their products. Further research needs to be performed in order to analyse the ways in which these firms gained these clients.

FUNDS AND OTHER FORMS OF SPONSORING

Early sponsoring

Access to financial resources is a critical problem for NTBFs and large established firms were sometimes approached for funding, either directly as equity shareholders or indirectly by being asked to provide advanced payments or support to early product development. However, on the whole, large established firms were found to play a limited role in the direct financial sponsorship of NTBFs. The incidence of equity shareholdings was very small among the NTBFs interviewed (a similar result being found in the larger sample surveyed). There were only 5 cases identified, all among the younger firms and all minority shareholdings, involving very small amounts of capital. Indeed, some NTBFs pointed out that the capital was the less relevant aspect of the shareholder's presence, since credibility and 'lobbying' power were more valued. Because venture capital was also rare, most NTBFs were launched on the basis of the founder's own resources, or else, resorted to bank loans or used government incentives to innovation as an indirect source of funding.

But there was evidence of other forms of large firm sponsorship, particularly among 3 categories of companies, comprising previous employers, equity shareholders and early clients. Table 12.3 presents some contributions of large 'sponsors' identified by the research. In some cases sponsorship was shared by other actors including research organisations, venture capitalists, incubators and other NTBFs.

Table 12.3: *Contributions of large sponsors to NTBF launch and early activity*

	Previous employer	**Equity shareholder**	**First client**
Finance	Provision of finance Facilitate access to other sources of capital	Start-up capital Credibility regarding further sources of capital	Advanced payments Steady orders
Technology	Access to external sources of technology	[Complementary skills and technology]	Customer requirements Sponsor development Test-bed for new products
Market	Demand for first product Integration into trade networks Contacts with potential clients References	Demand for first product Enhanced credibility Access to other clients Lobbying capacity	Trust in an unknown firm Pre-launch orders Custom expertise Credibility to other clients
Operational support	Operational support during launch	Advice for project definition [Managerial support]	
Complementary assets	Access to distribution channels	Manufacturing facilities Access to distribution channels	Access to distribution channels

These companies were found to have an important role in the launch and early life of NTBFs, either at the level of supply by providing tangible or intangible assets for the new firm, or at the level of demand through the direct or indirect provision of markets for its products. It was found that an association with reputed organisations facilitated access to external finance and, in turn, the presence of external capital eased market entry. The firms interviewed were especially emphatic about those who *actively* supported the process of early market entry by trusting an unknown firm, promoting the firms' activities or proposing clients.

Table 12.4: *Cases – examples of beneficial and adverse relationships with large firms*

Large partner that supports a difficult market penetration
The firm was the first to provide a sophisticated service at country level. Its launch involved a large investment and the founders were able to gain, as a shareholder, a large national company that needed the service (until then being forced to obtain it abroad) and thus became an important client from the start-up. The experience of similar firms in other countries led the new firm to target a market that in Portugal was not very receptive to the technology – although it was superior to the technology currently in use. Therefore, early survival depended on 'patient money' and continued demand from a large shareholder. A new and unforeseen market was eventually identified that allowed the firm to finally expand its client base.

Credibility and new clients through the ex-employer
The founders originated from a local subsidiary of a foreign MNE, which was a very important client at the time of formation. Informal contacts with the ex-employer (in Portugal and at other locations) and other contacts abroad, also obtained through the ex-employer, facilitated early technological problem solving. The credibility provided by working with a reputed company facilitated market entry, while 'word-of-mouth' about the firm's competence among the ex-employer's foreign customers and suppliers permitted the firm to gain some of them as new clients. This market expansion allowed diversification away from the ex-employer.

Lead-user and partner in internationalisation
The firm was launched in 1979, pioneering an emerging technology in Portugal. A public utility with extensive and diversified needs became an important client, providing a sustained demand. However this client was extremely unreliable for policy-related reasons, causing the firm to go through a serious financial crisis. After this first negative experience, the firm was able to gain another good client in a large technology aware company, that not only gave the NTBF an opportunity to develop a sophisticated product, but later became a partner in exporting the product by devoting its considerable resources and extensive commercial channels to this marketing effort.

Excessive dependence on lead-user and failure
A client with a substantial demand was the market opportunity identified by the firm. This client – a technology aware company – acted as a 'lead-user' supporting development and serving as test-bed for the new product. A close user-producer relationship was established and translated into further product developments. When interviewed, the founder noted the risks associated with a strong dependence upon one client, but although diversification was actively sought, the firm had not been able to gain clients which could replace the 'lead-user'. The final outcome confirmed his fears. In fact, when the client experienced a slow down in its market demand, this small supplier was forced to close down.

Acquisition in search for resources and 'close down'
This firm pioneered a new technology in Portugal in the early 1980s and experienced a troubled existence. It was nevertheless able to accumulate a reservoir of technological knowledge in its field of activity and reached a point when '*a move from the garage stage*' was envisaged. The need for resources to grow, and the opportunity to access complementary assets (at the level of production and marketing where deficiencies were acknowledged), led the firm to accept acquisition by a large technology intensive company. The founder expected to maintain some autonomy by integrating with a 'cluster' of firms with complementary skills that the large company was building. However, the buyer ended up closing down the small subsidiary, although keeping some of its employees.

Acquisitions of NTBFs by large companies

Large firms appeared to have a greater interest in acquiring NTBFs than being involved in start-ups. Among the NTBFs interviewed, 6 out of the 14 firms created before 1986 had been acquired by 1994. Several acquisitions were close to 'rescue' operations, caused by financial problems. But in other cases the objective of the NTBF was to obtain resources to promote growth, or to access missing competencies (e.g. at the level of marketing or production). However, most acquisitions were too recent to assess their outcome.

Indeed, there were some indications of 'predatory' behaviour on the part of a few large technology intensive firms operating in the electronics market and also of foreign MNEs operating in the software market. The acquisition (or attempt to acquire) of a number of NTBFs by these companies can be explained as a way of gaining quick access to NTBFs' technological competencies or knowledge of the Portuguese market (in the case of foreign firms), or as a way of absorbing potential competitors in markets where larger firms were seeking to become established. On the other hand, it is also possible to argue that some acquisitions carried out by large national companies favoured the creation of 'critical mass' in areas where size is a competitive factor. However, further research on these acquisitions is required to evaluate their effective outcome.

On the whole, although several NTBFs mentioned the importance of linkages with larger companies in providing access to non-technological assets, most of them did not favour partnerships. They were aware of their weaker position, and of the chances of being 'swallowed' by the more powerful partner. The example of the few firms which have been able to capitalise on their linkages with larger companies seems to suggest that a preferential, but *arms-length* relationship (where the small firm is important for the larger one, but does not depend on it), is the most favourable situation for a NTBF. However, this may not always be possible. Some firms exhibiting this type of relationship had already been approached to allow an equity shareholding by the large company. Most of these NTBFs had, so far, avoided such an option, but some of them were pessimistic regarding their ability to continue on their own.

Table 12.4 presents a number of examples, chosen from the group of firms interviewed, that illustrate cases where relationships with large firms had a positive impact on the survival and development of NTBFs and cases where the result of such relationship was not as effective as expected.

CONCLUSIONS

This research on the activities of NTBFs in Portugal must conclude that such firms have an important role acquiring, developing and diffusing new technologies into the country (Fontes, 1997). But it is also observed that, with respect to diffusion, the impact of NTBFs' activities are more evident at the level of customers that were technology aware and prepared to accept advanced solutions. A substantial proportion of these clients were large firms, but few of them were technology intensive or operated in sectors similar, or complementary to, the NTBFs'. The majority were large users in service sectors and in manufacturing industries other than electronics, which were technology informed and willing to upgrade technologically. This may explain why, with a few exceptions, the principal manifestation of interest of these companies in NTBFs technological capabilities was generated through customer-supplier relationships. The acquisition of NTBFs experiencing difficulties by larger more advanced companies was another way of accessing NTBF competencies. But the opportunities for technological collaboration at the industry level were scarce and they tended to occur between NTBFs. Therefore it can be argued that, with respect to technology, NTBFs have a greater influence upon local large established firms than the latter have upon NTBFs, both at formation and over time.

While having a limited role as source of complementary knowledge and technology, some large technology aware companies – both national companies and subsidiaries of foreign MNEs – were important for NTBFs at other levels. These firms were NTBFs' principal (sometimes sole) market. Often a large client was the 'market opportunity' identified by NTBF founders, and some of these clients behaved as 'lead-users', participating in the definition of custom requirements and providing a safe environment for testing and debugging the new product. Some NTBFs were also able to establish, over time, close customer-supplier relationships with technology aware clients. The sophisticated needs of the customer allowed the NTBF to develop more innovative products, a crucial issue when foreign market expansion was envisaged. Early sponsoring assisted the NTBF in the development of the new product, while a steady demand over time provided it with the incentive and the financial stability to achieve continued technology acquisition. Less tangible, but equally valued contributions, included enhancing the young firm's credibility, providing contacts to other potential clients, and facilitating access to manufacturing facilities and distribution channels.

But these relationships could lead to situations of great dependence that carried risks for the small firm. Dependence was often unavoidable at start-up, but it became particularly dangerous if the NTBF was unable to expand its market, or to replace an early good client when the client failed. Long-lasting user-supplier relationships could also present problems, since they may cause the firm to become locked to the specific needs of the larger client, reducing the scope of application of its products in other contexts and hence its capacity to react to changes in the client's demand. Occasionally NTBFs also experimented 'predatory' behaviour on the part of the large firm. This may explain why it transpired that although some market relationships evolved to partnerships (particularly in the case of foreign market expansion), NTBFs appeared to favour preferential, but arms-length, relationships.

It can be concluded that the absence of *large technology intensive companies* in Portugal precluded some of the 'complementarities' that are available to NTBFs in more advanced countries. However, a number of functions, often performed by these companies in non-technological areas, were provided in Portugal by *large technology aware companies*, that were interested in the knowledge and the products that NTBFs had to offer, since they were able to supply complementary assets the new firm did not possess. This type of 'complementarity' is also present in other countries. The difference with respect to Portugal was that, in most cases, it was the only one available and that the number of companies prepared to play these roles by complementing the activities of NTBFs and/or compensating for their weaknesses was much smaller than in more advanced countries.

The limited opportunities for technological collaboration, and the relative scarceness of opportunities for other types of complementarities to take place, were problematic to the small technology intensive firm. As becomes clear from the above discussion, demand-related problems emerged to be particularly serious. Only a small number of clients were prepared to act as 'lead-users' – or at least to buy complex new products from young national suppliers. The profound gap between an 'elite' of technology aware clients and the wider market, combined with the NTBFs' intrinsic weaknesses with respect to resources and market skills, caused several firms to encounter severe difficulties in surviving and/or growing. These problems appeared to be particularly severe in the biotechnology field, where not only technological competencies were almost completely absent at the industrial level, but established firms in other sectors revealed a low interest in biotechnology products or services, thus not providing enough demand to motivate firm creation.

Some firms were attempting to overcome these problems by engaging in foreign market expansion. However, market internationalisation requires a great deal of effort from a small company, takes time to produce results, and is usually easier to undertake when the firm already had some success in the domestic market. Therefore Portuguese NTBFs would benefit from the expansion of local demand. Government policies can greatly contribute to such expansion, either directly through public procurement – which encourages firms to go on developing sophisticated products and provide them with additional credibility – or indirectly through actions that stimulate demand for the technologies and their applications. As Rothwell (1994) points out, one effective way of supporting firms in the process of introducing new technologies is to provide the potential customers with the conditions in which their products might be used. Thus one form of gaining more clients to the products supplied by Portuguese NTBFs, particularly in the less advanced sectors, where Portugal specialises, is to devise policies which foster the technological awareness of such local firms, to encourage them to increase their technological level, and to prompt them to use local suppliers.

REFERENCES

Bahrami, H. and Evans, S. (1995) Flexible Re-Cycling and High-Technology Entrepreneurship, *California Management Review*, 37, 3, pp. 62–87.

Carlsson, B. and Jacobsson, S. (1991) What Makes the Automation Industry Strategic?, *Economics of Innovation and New Technology*, 1, pp. 257–69.

Cooper, A.C. (1985) The Role of Incubator Organisations in the Founding of Growth Oriented Firms, *Journal of Business Venturing*, 1, 1, pp. 75–86.

Coviello, N.E. and Munro, H.J. (1995) Growing the Entrepreneurial Firm. Networking for International Market Development, *European Journal of Marketing*, 27, 7, pp. 49–61.

Deniozos, D. (1994) Steps for the Introduction of Technology Management in Developing Economies: The Role of Public Governments, *Technovation*, 14, 3, pp. 197–203.

Dahlman, C. and Westphal, L. (1982) Technological Effort in Industrial Development – An Interpretative Survey of Recent Research, in Stewart, F. and James, J. (eds.) *The Economics of New Technology in Developing Countries*, pp. 105–37, Frances Pinter, London.

Fontes, M. (1997) Creation and Development of New Technology Based Firms in Peripheral Economies, in Klofsten, M. and Jones-Evans, D. (eds.) *Technology, Innovation and Enterprise – The European Experience*, Macmillan, London.

Fontes, M. and Coombs, R. (1995) New Technology-Based Firms and Technology Acquisition in Portugal: Firms' Adaptive Responses to a Less Favourable Environment, *Technovation*, 15, 8, pp. 497–510.

Fontes, M. and Coombs, R. (1996) New Technology-Based Firm Formation in a Less Developed Country: A Learning Process, *International Journal of Entrepreneurial Behaviour and Research*, 2, 2, pp. 82–101.

Fontes, M. and Coombs, R. (1997) The Coincidence of Technology and Market Objectives in the Internationalisation of New Technology Based Firms, *International Small Business Journal*, 15, 4.

Granstrand, O. and Sjolander, S. (1990) The Acquisition of Technology and Small Firms by Large Firms, *Journal of Economic Behaviour and Organisation*, 13, 3, pp. 367–86.

von Hippel, E. (1987) Cooperation Between Rivals: Informal Know-How Trading, *Research Policy*, 16, pp. 291–302.

Kenney, M. (1986) Schumpeterian Innovation and Entrepreneurs in Capitalism: A Case Study of the US Biotechnology Industry, *Research Policy*, 15, pp. 21–31.

Kim, L. and Dahlman, C.J. (1992) Technology Policy for Industrialisation: An Integrative Framework and Korea's Experience, *Research Policy*, 21, pp. 437–52.

Kulicke, M. and Krupp, H. (1987) The Formation, Relevance and Promotion of New Technology Based Firms, *Technovation*, 1, 6, pp. 47–56.

Laranja, M. (1995) Small Firm Entrepreneurial Innovation in Portugal: The Case of Electronics and Information Technologies, PhD Thesis, SPRU, University of Sussex.

Lawton Smith, H., Dickson, K. and Lloyd Smith, S. (1991) There Are Two Sides to Every Story: Innovation and Collaboration Within Networks of Large and Small Firms, *Research Policy*, 20, pp. 457–68.

Lundvall, B. (1988) Innovation as an Interactive Process: From User-Producer Interaction to National Systems of Innovation, in Dosi, G. et al. (eds.) *Technical Change and Economic Theory*, pp. 349–69, Pinter Publishers, London.

Monck, C.S., Porter, R.B., Quintas, P. and Storey, D.J. (1988) *Science Parks and the Growth of High Technology Firms*, Croom Helm, London.

Oakey, R.P. (1991) Innovation and the Management of Marketing in High Technology Small Firms, *Journal of Marketing Management*, 7, 4, pp. 343–56.

Oakey, R.P. and Cooper, S.Y. (1991) The Relationship Between Product Technology and Innovation Performance in High Technology Small Firms, *Technovation*, 11, 2, pp. 79–92.

OECD (1992) *Technology and the Economy: The Key Relationships*, Paris.

Olleros, F.J. and MacDonald, R.J. (1988) Strategic Alliances: Managing Complementarities to Capitalise in Emerging Technologies, *Technovation*, 7, 2, pp. 155–76.

Perez, C. and Soete, L. (1988) Catching Up in Technology: Entry Barriers and Windows of Opportunity, in Dosi, G. et al. (eds.) *Technical Change and Economic Theory*, pp. 458–79, Pinter Publishers, London.

Pisano, G.P. (1991) The Governance of Innovation: Vertical Integration and Collaborative Arrangements in the Biotechnology Industry, *Research Policy*, 20, pp. 237–49.

Pottier, C. (1988) Local Innovation and Large Firms Strategies in Europe, in Aydalot, P. and Keeble, D. (eds.) *High Technology Industry and Innovative Environments: The European Environment*, pp. 99–120, Routledge, London.

Roberts, E. (1990) Initial Capital for the New Technological Enterprise, *IEEE Transactions in Engineering Management*, 37, 2, pp. 81–94.

Roberts, E. (1991) The Technology Base of the New Enterprise, *Research Policy*, 20, pp. 283–98.

Roure, J.B. and Keeley, R.H. (1989) Comparison of Predicting Factors of Successful High Growth Technological Ventures in Europe and the USA, in Birley, S. (ed.) *European Entrepreneurship: Emerging Growth Companies*, Proceedings of the First Annual EFER Forum, Cranfield School of Management: EFER, pp. 189–222.

Rothwell, R. (1983) Innovation and Firm Size: A Case of Dynamic Complementarity; Or, Is Small Really So Beautiful?, *Journal of General Management*, 8, 3, pp. 5–25.

Rothwell, R. (1994) Issues in User-Producer Relations in the Innovation Process: The Role of Government, *International Journal of Technology Management*, 9, 5/6, pp. 629–49.

Rothwell, R. and Dodgson, M. (1991) External Linkages and Innovation in Small and Medium Sized Enterprises, *R&D Management*, 21, 2, pp. 125–37.

Shan, W. (1990) An Empirical Analysis of Organisational Strategies by Entrepreneurial High-Technology Firms, *Strategic Management Journal*, 11, 2, pp. 129–39.

Shearman, C. and Burrell, G. (1988) New Technology Based Firms and the Emergence of New Industries: Some Employment Implications, *New Technology, Work and Employment*, 3, 2, pp. 87–99.

Teece, D.J. (1986) Profiting from Technological Innovation: Implications for Integration, Collaboration, Licensing and Public Policy, *Research Policy*, 15, pp. 285–305.

Walsh, V. (1993) Demand, Public Markets and Innovation in Biotechnology, *Science and Public Policy*, 20, 3, pp. 138–56.

CHAPTER 13

David and Goliath: To Compete or to Sell?

G.M. PETER SWANN

INTRODUCTION

It is now quite well documented that some small high technology firms are particularly adept at radical innovation. New entrants seem to perform better than established firms in exploiting technological 'spillovers' from one industrial sector to another, and particularly those that arise from technological convergence. Sometimes, indeed, the new entrant is so successful in this respect that its technology becomes the *de facto* market standard, and it manages to sustain a long term competitive advantage on the back of this, and ultimately to become *the* major company in its market (e.g. Intel, Microsoft). In this sense, 'David succeeds in out-competing Goliath'. But it is often the case that the small firm, which develops a distinctive technological advantage in the short to medium term, finds that a large established competitor will 'catch up' in the longer run, and eventually drive the small firm out of an emerging market by market power, if nothing else. In this case, 'David knocks Goliath down, but not out, and ultimately Goliath recovers to assert himself.'

In this setting, the high technology small firm faces 2 obvious questions and a third more subtle question.

First, can Goliath be out-competed? Can a long term competitive advantage be sustained in this particular technology – and if so, how?

Second, if Goliath is 'down' but not 'out', when should David offer to sell Goliath his sling so that he (Goliath) can fight others? If the established firm shows signs of recovering from the initial challenge of the radical new technology, at what point should the small high technology firm aim to sell its technology to the established firm?

The third question is more subtle, since it has to be recognised that the market potential of a particular new technology is not exogenous or pre-determined. It depends, amongst other things, on the marketing investments made by producers of the new technology. It is sometimes argued that the personal computer would not have enjoyed the success it achieved had not IBM entered the market in December 1981. In terms of our metaphor, the total size of the market will be greater if Goliath recovers to compete in the market than if he does not. In the light of this the small high technology firm should also ask if it is better that David should aim at the start to sting Goliath, but not to knock him out, and to sell his sling at the best moment? Conversely, is it better that the small firm should from the start aim not to out-compete its large competitor but to sell its technology to the established firm? When does the price at which the small firm could sell its technology to the large exceed the value of the technology to the small firm on its own?

The aim of the paper is to begin an exploration of these questions within a simple analytical framework. The next section provides a few illustrative empirical examples to motivate the analysis that follows. The following section summarises some of the literature which suggests that, in some circumstances, small high technology firms are better placed to develop radical innovations than larger established firms. The paper then sets out some basic conditions under which the potential size of the market for a new technology is dependent on the perceived credibility of the suppliers. The following sections use this framework to estimate the different market outcomes that might be found if David Ltd decides to go it alone, or if it sells out to Goliath PLC, and to explore whether David is better to compete or sell out. The final section concludes with some suggestions for further development of these issues.

EMPIRICAL MOTIVATION[1]

The VHS/Betamax video cassette recorder case (Grindley, 1992) is sometimes quoted as an example of how superior network externalities allowed an inferior technology to win a standards race. In fact, as Grindley (1992) demonstrates, there are some other important lessons in the case, making it a highly relevant example of the sorts of issues discussed here. By rights, Sony would have been expected to win such a standards race, as they were the first (of the 2) to enter the market with their Betamax system, which enjoyed some technological superiority to the later VHS system introduced by JVC – although in one important respect the latter was superior. Moreover, Sony would count as an established firm, and also a firm capable of producing high quality technologies. JVC, on the other hand, was a smaller firm, which could not expect to win such a standards race alone, and was obliged, from the start, to engage in joint venture activity to make its challenge a reality. So while Sony adopted a closed position with their Betamax standard, and did not license other manufacturers to make it, JVC by contrast took an open position over its standard, allowing (and indeed depending on) other firms to produce to that standard. It is widely argued that this open strategy was decisive in establishing the VHS system as the *de facto* standard, in quite a short period of time.

While this story does not perhaps fit the simplest David and Goliath story outlined in the previous section, it does demonstrate how joint ventures between a smaller entrant, with a distinctive technology, and other companies with strong marketing and production credibility can succeed in races of this sort.

The contest between the Apple MAC and the PC (often called the IBM PC, though that company only accounts for a share of production) also illustrates these points well (Langlois, 1992). In this case, however, the 'David' of the story did not succeed because it persisted in trying to sell its own (possibly superior) standard, and was reluctant to 'sell out'. In addition, of course, Apple had entered the personal computer market before IBM, with its Apple II. But the market contest that has attracted most attention was between the established IBM PC standard and the Apple MAC that appeared some years later. As is well known, IBM adopted an open position with its standard, which was successful in helping to secure their PC as the *de facto* standard, even if it subsequently opened the way for low cost 'clone' competition. Apple, by contrast, took a closed position with respect to its standard, and has only recently licensed other firms to make Apple hardware. In retrospect, many commentators have seen this as a strategic mistake. The analogy with our story is a close one: when David found it impossible to 'out-compete Goliath' he should have 'sold', but he was reluctant to do so.

The PC applications software market provides many examples of small companies selling out to a larger established firm. It is widely recognised that, during the era of DOS-based applications software, an important part of the route to dominance in the spreadsheet software market involved nurturing as wide a range as possible of add-on and add-in products. These are pieces of software designed to run alongside the core software package, but to enhance its functionality in a number of ways (Swann, 1990; Shurmer and Swann, 1995). In the first instance, many (or most) of the add-ons and add-ins were produced by third party software houses, which were often very small software companies, with a very small range of products. This proliferation of add-on activity was found especially around the Lotus 1–2–3 spreadsheet package, and we have argued elsewhere (Swann, 1990) that this proliferation both resulted from, and reinforced, Lotus 1–2–3's position as the dominant spreadsheet software package during the DOS era. In some cases, these add-ons and add-ins were subsequently bought up by the leading software houses, and were sold, bundled, with the core package – rather than as an add-on.

In this example the 'David' firms are not directly competing with the established market leader, but rather they live in a symbiotic relationship to established companies. Nevertheless, the long term success of many add-on software producers has hinged on a timely 'sell'.

THE RELATIVE STRENGTHS OF SMALL AND LARGE FIRMS

This section, drawing on Swann and Gill (1993), summarises some of the key literature on why small firms may have an advantage in coping with radical technological change. While much economic theory tends to stress the phenomenon of persistent dominance, much of the literature on organisations, in contrast, is concerned with *organisational inertia*. But it is important to make a distinction here between *radical* and *incremental* innovation (Mansfield, 1968; Freeman, 1982). While a stream of incremental innovations can reinforce the position of existing market leaders (Nelson and Winter, 1982; Ettlie et al., 1984), radical innovations can present severe difficulties for established firms (Daft, 1982; Tushman and Anderson, 1986). The new and the small firm is often better adapted to create and exploit radical innovations (Rothwell and Zegveld, 1982).

While new entrants are often responsible for major innovations, many minor incremental innovations are introduced by existing producers (Gort and Klepper, 1982). Winter (1984) found that, in an *entrepreneurial* regime, new entrants were responsible for about twice as many innovations as incumbents, while in a *routinised* regime, established firms were responsible for the vast majority of innovations. As a result, it is common to find a net exit of firm numbers and increasing concentration when a technology matures (Mueller and Tilton, 1969; Gort and Klepper, 1982; Rogers, 1982).

A more recent and equally useful distinction is made between innovations that are *competence-destroying* and those that are *competence-enhancing* (Tushman and Anderson, 1986). Technologies that require fundamentally different competencies from those that the firm has developed to date can be classified as competence-destroying, while new technologies that can be mastered by firms using their existing technological competencies are competence-enhancing. Competence-destroying innovations are typically initiated by new firms, while competence-enhancing innovations typically emerge from existing firms. The difference between this distinction and the radical-incremental dichotomy is that competence-enhancing innovations need not be minor. In some cases, indeed, they can represent 'order of magnitude' improvements in a technology. But the main point is that competence-enhancing innovations do not render obsolete those skills which were used to develop the previous generation of technology.

Large incumbent firms often find it hard to cope with the challenges of competence-destroying innovations. Classic studies by Burns and Stalker (1961) and Stinchcombe (1965) suggest that such inertia may become embodied in an organisational structure. Burns and Stalker proposed the very influential distinction between *organic* and *mechanistic* organisations, where the former was better adapted to cope with rapidly changing conditions, while the latter is better adapted to stable and predictable market conditions, and indeed could exploit economies of scale in such a setting.

Nelson and Winter (1982) have argued that large complex organisations depend, in large measure, on tried and tested innovative routines, and are poor at improvising coordinated responses to novel situations. Routines can be seen as a truce in intra-organisational conflict: this is valuable and should not be broken lightly. When routines are in place, it is costly and risky to change them radically.

Studies in the area of the population ecology of firms (Hannan and Freeman, 1977; 1984) suggest a variety of reasons why established firms may exhibit structural inertia. These include internal pressures such as sunk costs, limits on the information received by decision makers, existing standards of procedure and task allocation. Moreover, since any organisational restructuring disturbs a political equilibrium, this may cause decision makers to delay reorganisation. For these and other reasons, established firms find it hard to adapt to a rapidly changing environment.

Nevertheless, many large firms have shown a great capacity to handle innovations that might have been competence-destroying (Mowery, 1983; Pavitt, 1991). A typical characteristic of these firms is that they are multi-divisional and operate over a broad range of technologies. They can be particularly successful in technologies which require a wide accumulated competence across disciplinary, functional and divisional boundaries (Teece, 1986; Pavitt, 1991).

Exploiting synergies depends on inter-divisional communication about different experiences (Mansfield, 1968), and may be best achieved by the exchange of personnel and experience (Aoki, 1986). Cohen and Levinthal (1989; 1990) point out that a firm's absorptive capacity does not merely refer to its interface with the external environment, but also its capacity to transfer knowledge between and within its own sub-units. An organisation's absorptive capacity does not depend so much on individuals, as on the network of linkages between individual capabilities (Nelson and Winter, 1982). And it is that learning and problem-solving, from diverse knowledge bases, that is most likely to produce innovation (Simon, 1985). To keep aware of new technological developments, and the competence to deal with them, firms must sustain their multiple technological bases (Dutton and Thomas, 1985).

NEW TECHNOLOGIES, CREDIBILITY OF PRODUCERS AND MARKET SIZE

This section sets out some simple analytical foundations from which to address the questions raised in the paper. The basic economic and strategic analysis of the *de facto* standards race provides the necessary building blocks for this analysis.

In the early analysis of standards (David, 1985; Farrell and Saloner, 1985; Katz and Shapiro, 1985) the race is between an established technology with a head start in the market and a loyal installed base of users who are unlikely to switch to a new technology, and a new entrant with a technically superior technology, but no community of users. It is assumed that buyers look at 2 generic features in making such purchase decisions. First, what is the intrinsic quality or performance of the technology? Second, how is the installed base of users of the technology likely to grow over time? The latter may be an important source of network externalities: these are indirect benefits that accrue to the purchaser of a technology that has a wide base of users. These could include the availability of parts, software, support and maintenance, and so on. In buying a technology with a well developed and loyal user base, the buyer can be reasonably confident that such benefits will continue. In contrast, buying a new technology, with no established user base, and with some uncertainty about how such a user base may grow in the future, and whether support will continue to be available, will inevitably appear to be a more risky investment.

In such models, where buyer behaviour is not co-ordinated, but instead buyers behave independently, it is quite likely that the market will stay loyal to an old technology. However, this outcome is not inevitable. One of the key issues for a firm trying to create a market for a new technology is to demonstrate its credibility, and its commitment to the market (Grindley, 1992). This can be done in a number of ways.

One general approach is for the firm to make investments in supporting the technology that are 'sunk costs': that is, they cannot easily be recouped if the firm abandons the technology. This signals to would-be users that the firm will have a strong incentive to stay with its technology so long as these sunk costs yield benefits, since when the technology is abandoned the investments are effectively lost. Examples of this generic strategy include investments in the development of user support materials that are specific to a new technology, investments in advertising, investments in productive capacity, and investments in developing 'add-on' products specific to the new technology.

Another approach is to build up a network of firms that supply the new technology, by adopting a fairly open stance towards licensing. By taking such an approach, the originator of the new technology may forego short term monopoly rents, but will more than compensate for this in terms of increased sales, if it can establish a winning position for its new technology in the technology race. A variant on this theme, especially relevant in the software market, is penetration pricing, where firms supply some versions of their software at a very low price (or even free) with the intention of building up the user base as rapidly as possible.

Thus to some extent it is in a firm's own hands how credible its new technology is perceived to be. Nevertheless, it is clear that the well established, large and profitable company must have a potential credibility advantage over the new, small firm. The history of the personal computer market from 1975 (Langlois, 1992) recognises that while there were some important precursors to the IBM PC, it was not until the entry of IBM into this market that many business users perceived the PC as a serious piece of office equipment. Certainly some aspects of IBM's entry strategy into the PC market (i.e. the open standard and the open approach towards licensing) can be seen in retrospect as playing an important role in ensuring the success of the PC standard (Grindley, 1992; Langlois, 1992). But the conclusion seems inescapable that the participation of IBM alone, regardless of IBM's strategy, played a most important role. Some speculations about what would have happened if IBM had not entered the PC market at that time (Langlois, 1992) consider that, while the entry of another 'major' into the PC would have given a boost to credibility, no other firm could have done so much for the future of the technology as IBM.

In short, the identities of the players *per se* are important factors in races of this sort. This observation is unremarkable, perhaps, but it plays an important role in the analysis of this paper.

Let us define the value that a particular user i derives at time t from a particular technology j as follows:

$$v_{ijt} = a_i Q_j + b_i S_{jt} + c_i N_{jt}$$

Where Q_j measures the quality or performance of technology j, S_{jt} is the level of support given to its technology by the producer of technology j at time t, and N_{jt} is the number of users of technology j at the start of the period t. This accumulated number of users depends, of course, on the history of purchase decisions up to time t.

In making investment decisions, however, the user is not just concerned about these factors today. He is also concerned about how they are likely to change in the future. For simplicity, we take the case where Q_j is fixed over time, and so neglects upgrade activity. (The results that follow, however, are not significantly affected by this assumption, although it renders the analysis simpler.) Nonetheless, it seems reasonable to assume that users will need to assess the growing profile of support over time (S_{jt}) and the growth of the user base N_{jt}, and it also seems likely that these perceived growth rates will depend on the identity of the producer of j. In terms of our metaphorical firms, users are likely to envisage a stronger growth path of S by Goliath PLC than for David Ltd, and this, in turn, would result in a faster growth rate of N. Suppose these perceived growth paths are as follows:

$S^e_{jt} = S_{j0} \exp\{g_j t\}$

$N^e_{jt} = N_{j0} \exp\{h_j t\}$

Where S_{j0} and N_{j0} are the levels of support and user base for technology j at time 0, and where g_j and h_j are the expected rates of growth of support and the user base for technology j. If we handle these expectations by assuming that buyers discount these streams at a rate of r over an indefinite future, we obtain the following expression for the user's gross present value of a technology (evaluated from period 0):

$V_{ij0} = a_i Q_j / r + b_i S_{j0} / (r - g_j) + c_i N_{j0} / (r - h_j)$

If V_{ij0} is greater than the price P_{j0} in period 0, then this technology would appear a viable purchase. The relationship implies that 3 things concern the buyer of a technology: its intrinsic quality, the expected path of producer support in future, and the expected growth path of the user base in future. The convenience of this formula, moreover, is that user choices are not sensitive so much to precise growth paths, but rather to the overall discounted weight of support and discounted weight of fellow users over the indefinite future. In what follows it is assumed that while David Ltd may have the superior technology, most users perceive that support from Goliath will grow faster, and possibly therefore that the user base will grow faster around the technology of the large established firm.

DAVID AND GOLIATH: THREE SCENARIOS

Shurmer and Swann (1995) show how a basic model of the sort set out in the last section can be developed to give a simulation model of the emergence of *de facto* standards with many competitors. Here we simply consider a much simpler set of 3 scenarios.

1. Goliath PLC continues to sell its established technology (only), while David introduces its superior technology in a purely competitive mode.
2. As (1) above, but with Goliath also planning a rival to David's technology.
3. David licenses Goliath to produce David's superior technology, and Goliath phases out its original technology.

While the precise numerical paths of sales for each technology in each scenario depends on several distributions and parameters, we can summarise the qualitative nature of these outcomes in a series of diagrams. The underlying assumption is that, at time 0 the buyer will choose the technology for which $V_j - P_j$ is largest; or, if this 'profit margin' is negative for all technologies, then no purchase is made. To simplify matters we normalise c_i at 1 for all buyers; this means that the value of increments to the network of

162 NEW TECHNOLOGY-BASED FIRMS IN THE 1990s

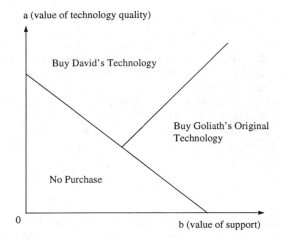

Figure 13.1
David competes with Goliath

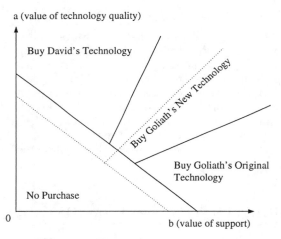

Figure 13.2
David competes with Goliath, old and new

users have the same absolute value to all users at all times. Figures 13.1 to 13.3, which map the purchase decisions of different buyers, summarise the outcomes under the 3 above scenarios, while Figure 13.4 compares the relative merits of solutions 1, 2 and 3 from the perspective of both David and Goliath.

Figure 13.1 shows the market outcome when David's new technology competes directly with Goliath's original technology. The Figure shows how the map of buyers with different values of **a** and **b** are divided between the 2 available technologies. It should be stressed that it will not be necessarily the case that buyers are distributed uniformly over this space, and hence relative areas on the map do not necessarily correspond to relative demand. Nevertheless, it is clear that David's new technology can cut deep into the market for Goliath's existing technology, especially amongst buyers who attach particular weight to the absolute quality of the technology, and less to the level of support offered by the producer.

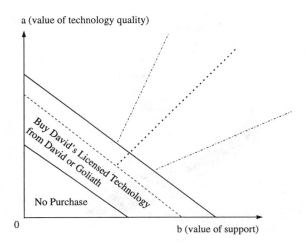

Figure 13.3
David licences Goliath

Figure 13.2 compares this outcome (see the dotted lines) with the outcome when in addition Goliath plans to introduce a version of the advanced technology. The picture, as drawn, relates to the position when Goliath has introduced the improved technology, although it is discounted back to the same period as represented in Figure 13.1. The assumption is that this new technology from Goliath is not of the same technical efficiency as that from David, but it is better supported. That is why this new technology attracts buyers from the middle of the map – in between those buying David's radical technology and those buying Goliath's original technology. But while this counter-move succeeds in winning back some of the buyer territory lost to David, the overall market sales are slightly reduced. This happens because the market is now more fragmented, being divided between 3 technologies as opposed to 2. The reduced network size attaching to each technology means that some buyers, who would previously have found it attractive to invest, no longer do so. From David's point of view, this second outcome is definitely inferior to the position included in Figure 13.1. But from Goliath's point of view, there is a fine balance between the extra territory gained from David, and the territory lost to 'No Purchase' in Figure 13.2.

Figure 13.3 compares these 2 outcomes with the solution when David 'sells' (or licenses) his technology to Goliath. Now the boundary between those who buy and those who do not moves down, and to the left. This happens because the joint venture between David and Goliath avoids the fragmentation of the market found above, so that the bulk of the expected network effects are concentrated in one technology, and this means that, at the margin, a greater number of buyers will enter the market. What Figure 13.3 leaves unresolved is the question of how these total sales are divided between the 2 producers. This arrangement will depend on the details of the licensing arrangement struck between David and Goliath. But what we can see is that it is highly likely that a mutually acceptable licensing arrangement can be found because the total market available to be shared between the 2 firms is very likely to exceed the sum of the maximum sales available to each party in any other setting.

Figure 13.4 illustrates this point. The best sales for David (as in Figure 13.1) are given by the area above the line DDXD. The best aggregate sales for Goliath are either those in Figure 13.1 – in which case it is clear that Figure 13.3 offers the potential of a net gain to both parties – or those in Figure 13.2, which in Figure 13.4 corresponds to

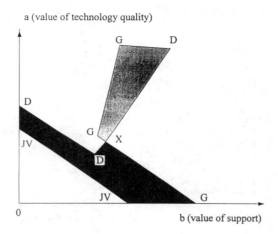

Figure 13.4
Comparison of strategies

the area to the right of GXGG. If the mass of demand in the trapezium GXDG in Figure 13.4 is outweighed by the mass of demand in the area DDXGJVJV, then the extra sales available to be shared between David and Goliath mean that both can do better in scenario 3 than in any other scenario. As drawn, the area DDXGJVJV is certainly larger than GXDG but, as we have noted above, the issue is also one of buyer density within this space. But given that the greatest density of buyers is likely to be close to the origin, and that buyer density drops off as we move away from the origin, then it is likely that we can find settings in which both are 'better off' in scenario 3.

IMPLICATIONS: COMPETE OR SELL?

In the simple model used above it is not possible for a David to 'out-compete' a Goliath completely, because even if David's radical new technology is considerably superior to that of Goliath, it is assumed that Goliath can eventually copy this technology to a comparable, if not exactly equal, standard, and because Goliath can offer support for its technology beyond that of David. The large firm may 'start from behind' in terms of the intrinsic quality of its technologies, but its entry to the advanced technology end of the market introduces a degree of credibility to the technology that was not present when there were only small scale entrants. The precise outcome depends, however, on the distribution of users in Figures 13.1 to 13.4, and in particular, on whether the majority of users give greater attention to technological quality or to producer support and the size of the established user base.

In the model, nevertheless, it is certainly possible for David to survive during a sustained market battle with Goliath. However, what the model illustrates is that it may be in the best interests of both parties that they form a joint venture, where David licenses Goliath to use his (David's) superior technology, and this joint venture of technological superiority and marketing strength serves to expand the market to the benefit of both.

It is suggested that an outcome of this sort may be very common indeed in markets where *de facto* standards are significant, and where it is important to users that they obtain direct support from producers and also the benefit of adopting a technology with a large user base. If such an outcome is so common, new technology small firms with a distinctive lead in a technology should have, from the outset, a contingency plan for the 'sell out' option. Rather than wait in a reactive fashion for a 'Goliath' suitor, the small firm may need to consider, from an early stage, which company would make an ideal partner, and when such a joint venture should be formed.

In the model, the virtue of the joint venture is that it expands the overall market for the technology. There were no other competitors (beside David and Goliath) and the *de facto* market standard was not yet resolved. This is what made the licensing joint venture an attractive proposition. But in a broader competitive setting there will only be a limited window of opportunity in which such a joint venture can hope to succeed in establishing a *de facto* standard. The length of this window is hard to estimate, even in retrospect, and especially in prospect! But as a first rule of thumb, the window may still be open when the existing user base is fragmented between several incompatible technologies, all of which are produced by small firms without the credibility and marketing muscle of an established large firm. In contrast, when the technology of another established firm has already achieved a status of the *de facto* standard, then the best that the small firm entrant with a superior technology can hope to achieve is to mark out a 'specialist niche' for itself (in the top left hand corner of Figures 13.1 to 13.4). In that case, a joint venture with a Goliath is more or less irrelevant.

CONCLUSION

This paper has explored the question of whether high technology small firms that enter markets possessing superior technologies can hope to compete over the long term against established companies with greater marketing muscle, who can bring greater credibility to their product technologies. We suggest that, in a range of settings, the small firm may not be able to out-compete its large counterpart, but can usefully secure its position in the market by licensing its superior technology to an established firm. Such a joint venture between technological competence and marketing credibility can make for a strongly competitive team in many high technology markets. Although this paper has presented an exploratory conceptual analysis of this phenomenon, we have not attempted to offer any detailed empirical analysis, or general rules of thumb. Nevertheless, we saw in a brief review of some empirical cases at the start of the paper that questions of competition arise in a variety of high technology markets. The model developed here is a very simple one, and it would benefit from more detailed elaboration in further research, to cover a wider range of more complex competitive environments. The longer term aim is to construct a simulation model of entry and joint venture strategies for the high technology small firm in such settings.

NOTE

1. The names of products and companies referred to in this paper are trademarks or registered trademarks of those companies.

REFERENCES

Aoki, M. (1986) Horizontal vs. Vertical Information Structure of the Firm, *American Economic Review*, 76, 5, pp. 971–83.
Burns, T. and Stalker, G. (1961) *The Management of Innovation*, Tavistock Publications, London.

Cohen, W.M. and Levinthal, D.A. (1989) Innovation and Learning: The Two Faces of R&D, *Economic Journal*, 99, pp. 569–96.

Cohen, W.M. and Levinthal, D.A. (1990) Absorptive Capacity: A New Perspective on Learning and Innovation, *Administrative Science Quarterly*, 35, 1, pp. 128–52.

Daft, R.L. (1982) Bureaucratic vs. Non-Bureaucratic Structure and the Process of Innovation and Change, in Bacharach, S.B. (ed.) *Research in the Sociology of Organisations*, 1, pp. 129–66, JAI Press, Greenwich, CT.

David, P. (1985) Clio and the Economics of QWERTY, *American Economic Review Proceedings*, 75, 2, pp. 332–6.

Dutton, J. and Thomas, A. (1985) Relating Technological Change and Learning by Doing, in Rosenbloom, R.D. (ed.) *Research on Technological Innovation, Management and Policy*, 2, pp. 187–224, JAI Press, Greenwich, CT.

Ettlie, J.E., Bridges, W.P. and Okeefe, R.D. (1984) Organisational Strategy and Structural Differences for Radical vs. Incremental Innovation, *Management Science*, 30, pp. 682–95.

Farrell, J. and Saloner, G. (1985) Standardization, Compatibility and Innovation, *RAND Journal of Economics*, 16, 1, pp. 70–83.

Freeman, C. (1982) *The Economics of Industrial Innovation*, 2nd Edition, Frances Pinter, London.

Gort, M. and Klepper, S. (1982) Time Paths in the Diffusion of Product Innovations, *Economic Journal*, 92, September, pp. 630–53.

Grindley, P. (1992) Standards, Business Strategy and Policy: A Casebook, Centre for Business Strategy, London Business School.

Hannan, M.T. and Freeman, J. (1977) The Population Ecology of Organisations, *American Journal of Sociology*, 83, pp. 929–64.

Hannan, M.T. and Freeman, J. (1984) Structural Inertia and Organisational Change, *American Sociological Review*, 49, pp. 149–64.

Katz, M. and Shapiro, C. (1985) Network Externalities, Competition and Compatibility, *American Economic Review*, 75, 3, pp. 424–40.

Langlois, R.N. (1992) External Economies and Economic Progress: The Case of the Microcomputer Industry, *Business History Review*, 66, pp. 1–50.

Mansfield, E. (1968) *The Economics of Technical Change*, Norton, New York.

Mowery, D.C. (1983) Industrial Research and Firm Size, Survival and Growth in American Manufacturing, 1921–46: An Assessment, *Journal of Economic History*, 43, pp. 953–80.

Mueller, D.C. and Tilton, J.E. (1969) Research and Development Costs as a Barrier to Entry, *Canadian Journal of Economics*, 2, 4, pp. 570–9.

Nelson, R.R. and Winter, S.G. (1982) *An Evolutionary Theory of Economic Change*, Harvard University Press, Cambridge, MA.

Pavitt, K. (1991) Key Characteristics of the Large Innovating Firm, *British Journal of Management*, 2, 1, pp. 41–50.

Rogers, E.M. (1982) Information Exchange and Technological Innovation, in Sahal, D. (ed.) *The Transfer and Utilisation of Technical Knowledge*, D.C. Heath, Lexington, MA.

Rothwell, R. and Zegveld, W. (1982) *Innovation and the Small and Medium Sized Firm*, Frances Pinter, London.

Shurmer, M. and Swann, P. (1995) An Analysis of the Process Generating De Facto Standard in the PC Spreadsheet Software Market, *Journal of Evolutionary Economics*, 5, pp. 119–32.

Simon, H. (1985) What do we Know About the Creative Process?, in Kuhn, R.L. (ed.) *Frontiers in Creative and Innovative Management*, pp. 3–20, Ballinger, Cambridge, MA.

Stinchcombe, A.L. (1965) Social Structure and Organisations, in March, J.G. (ed.) *Handbook of Organisations*, pp. 153–93, Rand McNally, Chicago.

Swann, P. (1990) The Growth of a Software Network, in Berg, J. and Schumny, H. (eds.) *An Analysis of the IT Standardisation Process*, North Holland/Elsevier Science Publishers, Amsterdam.

Swann, P. and Gill, J. (1993) *Corporate Vision and Rapid Technological Change*, Routledge, London.

Teece, D. (1986) Profiting from Technological Innovation: Implications for Integration, Collaboration, Licensing and Public Policy, *Research Policy*, 15, 6, pp. 285–305.

Tushman, M.L. and Anderson, P. (1986) Technological Discontinuities and Organisational Environments, *Administrative Science Quarterly*, 31, 3, September, pp. 439–65.

Winter, S. (1984) Schumpeterian Competition in Alternative Technological Regimes, *Journal of Economic Behaviour and Organisation*, 5, 3/4, September/December, pp. 287–320.

PART VII Regional Factors and the HTSF

CHAPTER 14

Knowledge Co-operation in the Dutch Pharmaceutical Industry: Do Regions Matter?

MARINA VAN GEENHUIZEN

SETTING THE SCENE

Regional clusters of industry have attracted considerable attention in regional economic analysis in recent years. The main arguments proposed in stressing the importance of regional clusters are based on a generally accepted shift from mass production to product differentiation and fast moving consumer needs. This shift is also connected with new production modes organised on a flexible basis near to the market.

Some authors have taken a shift to flexibility in production for granted. Others have challenged the pervasive nature of this phenomenon (e.g. Gertler, 1988) and advocated a simultaneous rise of alternative organisational forms. Various opposing opinions have also been forwarded on the spatial impact of flexible modes of production. On the one hand, it has been speculated that a reconcentration of production is taking place (e.g. Schoenberger, 1987; Storper & Walker, 1989). In this view spatial agglomeration is associated with a trend for vertically disintegrated and locationally fixed production (Scott, 1988; Storper & Scott, 1989). On the other hand, many researchers have pointed to a need for sensitivity to spatial diversity in the analysis of the organisation of production (e.g. Gertler, 1988; Amin, 1993; van Geenhuizen and van der Knaap, 1994; Saxenian, 1994). Such spatial diversity manifests itself in the spatial scale of networks (both local and global) and the type of relationships involved.

There is one essential reason why the pharmaceutical industry calls for the above sensitivity toward spatial diversity. A shift to flexible manufacturing would not be appropriate in this industry. There is no influence of fast changing fashion and product differentiation. Patent protection stabilises the use of particular technologies for a certain number of years. What is important in the pharmaceutical industry is an increasingly expensive R&D, causing a need for out-sourcing of research to universities and other research institutes (Howells, 1992; Ministry of Economic Affairs et al., 1993). In view of new drug developments based on biotechnology, research co-operation with (small) biotechnology firms is equally important. There is also a strong need for networking during marketing. Accordingly, the pharmaceutical industry derives benefits from networking, mainly in non-production activities, rather than from the production process itself.

There is also a need for a differentiated approach towards research and knowledge creation between various high-technology industries. Innovative pharmaceutical companies are often highly specialised in a few therapeutic classes, such as anti-hypertensia, anti-ulcer and antibiotics, based upon increasingly expensive R&D. These properties lead to limited knowledge flows between pharmaceutical companies, and a dominance of formal communication (e.g. Storper, 1996).

This chapter focuses on the pharmaceutical industry in the Netherlands and explores the role of the regional environment with regard to co-operation to produce knowledge. The Dutch pharmaceutical industry is a small, stable and research-intensive sector. There has been a recent increase in employment of 16% between 1980 and 1993, a pattern that is clearly different from total manufacturing (i.e. a decline of 19%). In addition production, in terms of turnover, has grown at a significantly higher rate than total manufacturing (i.e. 148% versus 36%). Currently R&D expenses amount to 9% of turnover in this sector, which is well above the national average (van Geenhuizen and van der Knaap, 1997).

This chapter will first discuss various forms of learning processes by companies associated with innovation, with a particular focus on the modes of communication involved, and implications for the distances experienced. This is followed by a reflection on theoretical concepts of regional networks and an empirical exploration of the relevance of such networks in past innovation in the Dutch pharmaceutical industry. Furthermore, present-day needs for knowledge co-operation, particularly out-sourcing, are given attention with an emphasis on the spatial scale of the relationships involved. This includes also the relationships with small biotechnology firms, as a source of knowledge, basically about new drugs. The chapter concludes with various policy implications.

Evidence on the competencies of companies and their regions is derived from desk research (using annual reports of companies and company life-histories), as well as interviews by telephone and in-depth interviews with management at production sites. In sum, the evidence draws on 17 pharmaceutical firms, the 4 largest covering approximately half of the employment in the sector, and 10 biotechnology firms.

PATTERNS OF LEARNING AND COMMUNICATION

Learning has increasingly achieved attention as a major activity for companies during the execution of their strategies for survival and growth (e.g. Dosi et al., 1990; Camagni, 1991; Malerba, 1992). Stimuli to learning come from developments in competition, e.g. new entrants and substitution, government regulation, new technologies (particularly pervasive ones such as information technology, new materials and biotechnology) and new modes of production organisation.

Much knowledge needed for innovation draws on the companies' continuous in-house learning-by-doing and learning-by-using activities which are connected with routine activities (Malerba, 1992). In addition, in-house R&D (learning-by-searching) makes 2 types of input available to the companies' knowledge base. First, there is new knowledge from planned research (but also from unexpected sideways research paths and failures). Secondly, there is an enhancement of the learning capability, i.e. the ability to detect, assimilate and exploit developments in technology in the external environment, especially when these developments are near to the existing expertise of the R&D staff. Knowledge is also drawn from the external environment by interacting with suppliers, users and research institutes (learning-by-interacting), and by monitoring what competitors are doing. In recent decades companies have tended to rely to a greater extent on external knowledge sources.

Quite recently it has been emphasised that learning, knowledge development and concomitant communication differ significantly between economic sub-sectors and product categories (Storper, 1996). For example, there is a clear difference between dedicated products by specialised producers and generic specialised high-tech products. In the former, learning takes place as a result of an ongoing interaction between producers and users of the technology (learning-by-interacting). Accordingly, communities of specialists design and redesign products over very short time horizons by using tacit and customary knowledge, with continuous communication as an important condition for carrying out this type of innovation. Such communities are typically concentrated in particular geographical areas where informal communication is critical to successful operation. Examples are craft-oriented industries such as in the Third Italy, but also non-merchant semi-conductor production such as in Silicon Valley (Saxenian, 1994).

With regard to generic specialised high-tech products, innovation is based upon the use of highly codified scientific (engineering) knowledge and formal processes of R&D. This involves forms of communication that can be planned at regular time intervals (through meetings and congresses) and stretched over long distances. However, at the same time, innovation is dependent on the previously mentioned informal and interpersonal processes for some of the cutting-edge knowledge inputs. Where uncertainty exists, there is a need for informal interaction in unplanned and uncodified ways (Storper, 1996). This pattern of mixed formal and informal learning and communication is exemplified by the innovative pharmaceutical industry, with different needs in the various stages of R&D, with biotechnology as one of the cutting-edge inputs.

In the late 1970s, when biotechnology entered the commercialisation stage, most established pharmaceutical companies had no in-house knowledge in the field of molecular biology, the discipline underlying genetic engineering. In fact, small biotechnology firms appeared to be excellent partners for knowledge co-operation with large established pharmaceutical companies in order to fill this knowledge gap (Sapienza, 1989; Arora & Gambardella, 1990; Senker, 1996). Small biotechnology firms have usually the flexibility and creativity to generate new ideas that cross borders, whereas large established firms have formalised procedures and clear defined research areas and boundaries (Daly, 1985; Howells, 1992). Presently, biotechnology firms are promising sources of new drugs based upon their specific knowledge of the human genes (Kohler, 1996).

New knowledge, particularly new product technology, is the major input in the innovative pharmaceutical industry, because it is the single most important weapon in competition (Sapienza, 1989; Anderson, 1993; van Geenhuizen & van der Knaap, 1997; Halliday et al., 1997). Basic R&D in the pharmaceutical industry has become increasingly expensive over the past years. On average, it takes 12 years of R&D and an investment of 500 million DFL before a new drug can be introduced to the market (Nederlandse Vereniging van de Innoverende Farmacentische Industrie, 1995).

Various unique circumstances in the industry have led to high costs and high risks of R&D since the early 1980s (Ministry of Economic Affairs et al., 1993; van Geenhuizen & van der Knaap, 1997). First, conventional research areas (paths) became depleted so that entirely new ones had to be opened, with concomitant risks and failure of new drugs. Secondly, registration procedures have increased in tightness and complexity, as well as duration. These regulatory requirements have shortened the effective life of patents, i.e. protection while actually on the market. But the situation has improved since 1991 due to new patent legislation in the European Union extending patent duration. Third, in order to reduce costs of health care services, many governments (including the Dutch) have promoted the prescription of generic drugs, i.e. low-priced chemically identical copies of drugs of which the patent period has expired. This policy has hampered the basically innovative pharmaceuticals industry from gaining a sufficient level of return-on-investment of R&D, a situation which has reinforced the need for collaboration and the out-sourcing of R&D.

Currently, a major new threat is increased competition within the industry. The period of an exclusive position of new drugs in the market becomes shorter and shorter by the action of competitors (Kohler, 1996). This makes collaboration with biotechnology firms as a source of new drugs even more pressing.

REGIONAL NETWORKS AND INNOVATION IN THE PAST

Since the mid 1980s, many old and new network concepts have gained popularity in the analysis of spatial concentration of economic activity, such as complexes, growth poles, and clusters (Håkansson, 1988; Boekema & Kamann, 1989; Camagni, 1991; Lambooy, 1991). A very generic concept is a network, meaning a set of interdependent organisations (actors). Networks have been recognised as an important organising principle for interaction between companies. This thinking stems from a renewed interest in the work of Coase (1937) on firms and transaction costs, new ideas on intermediate links between firms and markets (Richardson, 1972; Williamson, 1985), and the recognition of a growing static and dynamic uncertainty in corporate reality (Camagni, 1991). The basic assumption of network relationships as a third type of organisation beside market and hierarchy types is that one actor is dependent on resources controlled by other actors, and that benefits can be gained by the pooling of resources (Powell, 1990).

In the analysis of corporate networks many structural elements are generally taken into consideration (Håkansson, 1988; Håkansson & Johanson, 1993; Kamann, 1993). Five of them can perhaps be counted as the most important, namely:

1. actors such as suppliers, clients, and competitors,
2. resources such as human capital and new technology,
3. activities such as transformation, transaction, and innovation,
4. relationships between the actors,
5. channels (physical and non-physical) that facilitate interaction such as roads and share-holdings.

With regard to the competitive advantage and growth achieved by networking, the following properties of relationships are considered most important: the strategic content (such as cost economising, capturing of new knowledge), the spatial dimension (density, scale), the time dimension (permanent or temporary), the resources and activities involved, and the distribution of power (control and dependency). Accordingly, diverse concepts of regional networks have been developed with a different emphasis on the nature of the relationships involved (Lambooy, 1991; Wever, 1994; van Geenhuizen & van der Knaap, 1997):

- Complexes: dense functional relationships within a relatively small area, based upon one or more basic industries (Chardonnet),
- Growth poles: functional relationships as carriers of growth with one large, dynamic firm generating economic growth in other firms (polarisation effects) (Perroux),
- Filières: functional relationships between suppliers and purchasers (economically and technically) in all relevant directions, from raw materials or components to final products,
- Clusters of industry: institutionalised supportive relationships between an industry and skilled suppliers, customers, information networks, and public bodies, with external economies for all partners (Porter, 1990).

The spatial configuration of the pharmaceutical industry in the Netherlands is dominated by a relatively large concentration in the Core area (i.e. the Randstad) at the end of the 1980s (Table 14.1). This part of the Netherlands accounts for 58% of employment in the pharmaceutical sector, versus 46% of total employment (manufacturing and services). The majority of jobs here are in the sub-regional grouping of Delft and Westland (22%), and Gooi and Vecht (16%). Furthermore, small concentrations can be observed in the sub-regions of Utrecht and Agglomeration Haarlem (each with 6% of all employment). In the Intermediate Zone there is one large concentration in the region of Northeast North-Brabant (22%). This spatial pattern stems largely from economic specialisation already present in the 1920s and 1930s, particularly in the food industry. Good examples are Gist-Brocades in Delft (penicillin, emerged from alcohol and yeast production) and Organon (presently AKZO) in Northeast North-Brabant (hormones, derived from meat processing). In the remaining section, the industry in 2 regions, Northeast North-Brabant and the northern Core area (including Haarlem, Amsterdam, Gooi- and Vecht-region) (see Figure 14.1) will be considered in detail with regard to the above concepts of regional networks, with particular concern for knowledge co-operation at the regional level.

Table 14.1: *Regional distribution of employment in the Dutch pharmaceutical industry in the late 1980s (percentage share)*[1]

Regions	Pharmaceutical Industry	Total Industry and Services
Core Region	58	46
Delft Westland	22	
Gooi- and Vecht-region	16	
Utrecht	6	
Agglomeration Haarlem	6	
Intermediate Zone	33	26
NE North-Brabant	22	
Outer Zone	9	28
Totals	100	100

[1] Sub-regions with less than 5% of jobs excluded.
Source: Adapted from van Geenhuizen (1993).

The concentration of pharmaceutical industry in Northeast North-Brabant can be described as a small complex. It includes almost all stages of the product sequence and is dominated by dense functional relationships between the various firms. Knowledge co-operation between local firms has been important here, but it was mainly based on intra-firm links. The 'leader' company was (and is) Organon which expanded in the region over time, by absorbing Nobilis (later Interveth) (1961) and founding Organon Teknika (1971). The importance of knowledge co-operation is clearly evident in the synergy between R&D at Organon and R&D at particular subsidiaries (Tausk, 1978). For example, the technology on human reproduction hormones developed by Organon was subsequently successfully applied in animal drugs by Nobilis. Whether this regional intra-concern network also gave (and continues to give) access to outsiders in the region remains questionable. Supported by other studies of the region (van Geenhuizen, 1993), it can be stated that the network was (and is) relatively closed to other regional companies. The most obvious explanation for this phenomenon is a relatively high observed secrecy in product technology development.

It must be added that, simultaneous with the development of the regional network, Organon (later AKZO Pharma) continued to establish a global knowledge network.

Particularly in the 1980s, AKZO Pharma (and subsidiary Organon Teknika) were active in capturing knowledge in the US by direct investment in high-technology companies.

In contrast to Northeast North-Brabant, the industry in the northern core area is composed of many large and smaller firms belonging to a wide range of corporations. There are a few American established subsidiaries (e.g. Merck Sharp & Dohme, Centocor), aside from various indigenous firms that have been acquired by different (often foreign) companies in the 1970s and 1980s, e.g. Laboratoire Sarget (France), Byk Gulden (Germany), Solvay (Belgium), Medicopharma (Dutch) and OPG (Dutch). Given the strength of intra-firm relationships in R&D, it is reasonable to assume that networking towards the exchange of new knowledge has never developed within the region. It needs to be added that almost all firms were focused on their own highly specialised product technologies, a situation in which collaborative links aimed at knowledge exchange offered no advantages. At the same time, the companies in this region could (potentially) benefit from a supportive knowledge infrastructure providing access to local knowledge and global knowledge. This infrastructure includes academic hospitals in Leyden, Amsterdam (2) and Utrecht, the Dutch Cancer Institute and the Central Laboratory for Blood Products (both in Amsterdam), the National Institute for Health and Environment (near Utrecht), and the Central Veterinary Institute in Lelystad.

Given the above considerations, the industry in the northern core area cannot be classified according to the previously indicated concepts of regional networks. It is merely a 'formation' (Kramer, 1991) because of the (potential) common use of the regional production environment as the major (and often only) relationship between the companies.

It can be concluded that inter-firm co-operation in the pharmaceutical industry aimed at the capturing of knowledge has been almost absent on a regional basis, unless the companies belonged to one and the same (large) corporation. The next sections will focus in detail on modern knowledge co-operation, particularly the phenomenon of out-sourcing of R&D and collaboration with small biotechnology firms.

MODERN KNOWLEDGE CO-OPERATION AND THE REGION

In recent times, increasingly expensive R&D, coupled with a reinforced competition, has advanced the strategy of cost reduction of in-house R&D in the innovative pharmaceutical industry. This is evident in the growth in out-sourcing (Howells, 1992; Ministry of Economic Affairs et al., 1993).

Out-sourcing is a mode of technology co-operation with a relatively low interdependence between the partners involved (Hagedoorn, 1993). In general, one can distinguish between various forms of organisational interdependence, with joint ventures and research corporations at one extreme (i.e. large interdependence) and mutual technology exchange agreements and one directional technology flow (out-sourcing) at the other extreme (i.e. both small interdependencies). In innovative manufacturing, there seems to be a general trend towards short term, flexible and less dependent modes of knowledge co-operation (Geenhuizen et al., 1996), but it is reasonable to assume that in the pharmaceutical industry with its highly specialised and often secret knowledge, the relationships are long-lasting and focused on a few partners.

What would be the potential role of regions in the survival and profitability of the modern pharmaceutical industry given the forms of learning and communication that dominate in this industry? There is increasing evidence that only specific regional environments are able to adapt quickly to the high levels of uncertainty that surround knowledge-intensive industries (Andersson, 1991). The ability of regional environments to respond to needs of knowledge-intensive companies depends upon 3 essential activities (Geenhuizen et al., 1996), i.e. (1) the generation of new knowledge, (2) education and training, and (3) various networking activities such as in the transfer of knowledge. These characteristics are emphasised in modern management literature (Kanter, 1995) with regard to the needs of 'world class' companies. 'World class' companies preferably locate in places providing one (or more) of the so-called Cs comprising: concepts, competencies and connections.

Concepts are concerned with the latest knowledge and innovative ideas. The creation of new knowledge occurs in a well-structured form in universities, research institutes, and R&D departments in companies. As previously indicated, new knowledge is also the result of unexpected events and casual meetings, where informal procedures and uncodified language dominate. Some regions are superior environments in stimulating the latter type of knowledge creation, i.e. by creativity and serendipity among innovative people. The second resource – competence – refers to the ability to conform to the highest standard available. Some regions are superior sites by maintaining high quality standards and strong investment in skills and education. This includes not only academic knowledge, but also practical technical skills and management abilities required to implement innovations. Expertise and skills in the labour market are the most important local asset for innovative industry in Europe (Drewett et al., 1992).

The third resource (i.e. connections) is concerned with relationships that provide access to resources around the world, particularly gateways to global knowledge networks. Connections are also essential in the transfer of knowledge from creator to user (Charles & Howells, 1992) and in providing a sufficient match between the demand for knowledge and the supply of knowledge.

PATTERNS OF R&D OUT-SOURCING

In general, pharmaceutical companies can adopt 2 strategies in order to cope with increased competition. First, there is a defensive strategy in which their position in care management is strengthened by acquisitions or by coupling their interests with insurance firms, care companies, and drug trading companies. Second, there is an offensive strategy in which their innovative capacity is strengthened. The latter strategy includes out-sourcing of R&D to universities and research institutes, and an increased collaboration with small biotechnology firms.

In 1991, 30% of total R&D expenditure of the Dutch pharmaceutical industry was already concerned with bought-in R&D (Nederlandse Vereniging van de Innoverende Farmacentische Industrie, 1995). The trend for out-sourcing is also reflected in a decline of R&D employment since 1988. The number of R&D jobs has dropped by 11% between 1988 and 1994. In addition, as a share of total employment, there has been a downward trend, for example, from 19.2 to 17.3% between 1988 and 1993 (Table 14.2).

Out-sourcing of R&D needs a careful selection of partners, to produce different outcomes at each of the following stages of R&D (Ministry of Economic Affairs et al., 1993; Nederlandse Vereniging van de Innoverende Farmacentische Industrie, 1994):

1. discovery and synthesis of new chemical entities (NCEs),
2. pre-clinical testing (on animals) aimed to determine impacts, safety and quality,
3. clinical testing of new drugs on human beings,
4. technical processing, which (if successful) leads to

Table 14.2: *Employment in R&D in the Dutch pharmaceutical industry*

Year	R&D Employment	
	Number of Jobs	Share in Total Jobs (%)
1983	1.790	15.7
1984	1.820	15.7
1985	2.050	17.2
1986	2.110	17.1
1987	2.320	18.1
1988	2.480	19.2
1989	2.350	17.7
1990	2.350	17.7
1991	2.220	16.9
1992	2.300	17.3
1993	2.300	17.3
1994[1]	2.200	16.9

[1]Tentative figures.
Source: *Adapted from CBS (in Nederlandse Vereniging van de Innoverende Farmacentische Industrie, 1995).*

5. the start of a registration procedure, and Post-Marketing Surveillance (PMS), including additional investigation and monitoring of patients after market introduction.

Generally, the focus of knowledge co-operation in the pharmaceutical industry has become increasingly global. It appears nevertheless that, when considering R&D in detail, the spatial scale preferred is smaller (i.e. local or regional) provided that the right (highly specialised) expertise is available (Table 14.3). From the viewpoint of the outsourcing company the spatial orientation is preferably local/regional in 3 stages, i.e. discovery and synthesis of NCEs, pre-clinical testing, and technical processing. Critical factors in the selection of research partners are world-wide recognised expertise including a high rank in the citation index (the discovery and synthesis of NCEs stage), careful methods of clinical practice, and favourable legislation (the pre-clinical tests stage), and presence of a fine-chemical industry with expertise in process technology and chemical technology (the technical processing stage).

The spatial orientation is much wider (international) in clinical testing and in post-marketing surveillance (Table 14.3). Critical factors in partner selection here are largely associated with the quality and infrastructure of the medical system, educational level of physicians, and language and culture (the clinical testing stage), and the infrastructure of pharmacists and their registration systems (the post-marketing surveillance stage). It must be emphasised that the Netherlands holds an excellent international position in facilities for clinical testing.

Based upon the above orientation in out-sourcing, it can be concluded that the Dutch pharmaceutical industry (potentially) benefits from regional knowledge sources in selected R&D stages, due to availability of the precise expertise. In addition, various foreign pharmaceutical industries make use of R&D in the Netherlands in connection with clinical testing.

In empirical research on out-sourcing it has become evident that conditions for networking deserve to be improved in the Netherlands. The following obstacles call for attention and are currently being removed (Ministry of Economic Affairs et al., 1993; Nederlandse Vereniging van de Innoverende Farmacentische Industrie, 1994). First, there is a shortage of organised collaboration between industry, universities and research institutes, a situation that has led to a gap between basic research areas and

Table 14.3: *Selection criteria for partners in R&D out-sourcing*

Stage	Spatial Orientation	Critical Factors
1. Discovery/synthesis NCEs	Local/Regional	• Expertise
2. Pre-clinical tests	Local/Regional	• Methods (GCP)[1] • Legislation
3. Clinical tests	International	• Medical Infrastructure • Educational Level • Language and culture • Facilities
4. Technical processing	Local/Regional	• Presence of fine-chemical industry • Expertise
5. Registration	Internal (pharma companies)	–
6. PMS	International	• Pharmacists infrastructure

[1] Good Clinical Practice rules
Source: Van Geenhuizen & van der Knaap (1997)

applied (clinical) research. There is therefore a need for the establishment of an effective interface between industry and universities contributing to a better match between the demand and supply of knowledge. Secondly, the Netherlands is too small to excel in a wide variety of topics. Accordingly, there is a need for a narrower focus on basic research, such as on cardiovascular diseases and endocrinology. A final point is concerned with the training program of physicians with regard to clinical testing. More attention needs to be given to the specific needs of the pharmaceutical industry, clinical research and pharmacotherapy (Ministry of Economic Affairs et al., 1993).

COLLABORATION WITH BIOTECHNOLOGY FIRMS

In the Netherlands, around 35 biotechnology firms were active in R&D or manufacturing of pharmaceutical compounds, drugs or diagnostics in the mid 1990s. Their origins are rather diverse but they conform to what is observed in other countries (Daly, 1985; Haug, 1995), i.e. such new enterprises emanate from mature pharmaceutical firms specialised in traditional fermentation and enzyme production, other basically innovative pharmaceutical firms, and American biotechnology firms expanding through subsidiaries in Europe. They are also established as spin-offs from universities and research institutes, and spin-offs from other industries such as pharmaceuticals and chemicals. Related with this diversity of origin is a diversity in networking within the pharmaceutical industry.

Established subsidiaries of pharmaceutical companies are intended to perform basic research in inter-disciplinary fields in an informal setting, which in a formal organisation would be otherwise impossible. Dependent on their high innovative level, these new biotechnology firms tend to act on a global scale. Established subsidiaries of American biotechnology firms have a two-sided mission, namely first, to enter the European market and to conduct applied R&D in order to satisfy European rules of registration, and second, to connect with European (including Dutch) knowledge networks of universities, academic hospitals and research institutes, with a focus on almost all stages of R&D. These companies seldom enter into knowledge networks with the Dutch pharmaceutical industry.

Small spin-off firms, on the other hand, have a strong need for collaboration with the pharmaceutical industry, particularly after some years of existence when a shortage of resources for basic R&D and marketing becomes evident (Daly, 1985; Deeds and Hill, 1996; Senker, 1996). In addition, small biotechnology firms lack the competence of 'institutional engineering', necessary to negotiate with regulatory authorities and the medical profession (Walsh, 1996). Preliminary evidence on the spatial scale of collaboration of small spin-offs indicates a preference for partners within the region and the country. This pattern may be associated with the fact that most of these firms have a need for knowledge (channels) concerning the home market via partners within the same institutional setting as themselves. Although operating in highly innovative areas, small spin-off firms are often not first-innovators, but merely early followers. Only a few of these firms are first-innovators and they tend to operate in global knowledge networks.

Seen in terms of the Dutch pharmaceutical industry, knowledge networking is primarily based upon the availability of highly specialised expertise. If the right expertise is available in the region, networking on a regional basis has advantages above networking over longer distances.

POLICY IMPLICATIONS

This paper has explored the role of knowledge networking in the pharmaceutical industry in the Netherlands, by focusing on networks within the pharmaceutical industry, R&D out-sourcing by the industry to universities and research institutes, and networking by the industry with biotechnology firms. The evidence has indicated a small role for knowledge co-operation on a regional scale between pharmaceutical companies, and between pharmaceutical industry and small biotechnology firms. Two factors prevent the emergence of strong collaborative links between regional companies, namely highly specialised expertise and secrecy in new product technology. Regions may, however, play an important role in providing facilities for out-sourcing of R&D to universities and research institutes, both for regional companies and companies located somewhere else, provided that there is a sufficient match between demand for knowledge and supply of knowledge.

Quite recently, the Dutch government has adopted a general policy for improving conditions for knowledge-intensive industries (Ministry of Economic Affairs et al., 1995), based upon the recognition that the Dutch economy has lost strength in this respect in the past years (Sociaal-Economisch Raad, 1995). The new knowledge policy includes financial measures seeking to advance R&D by companies, and the allocation of additional budgets to advance top quality research and modernise vocational training. In addition, collaboration between industry and universities and other public research institutes will be enhanced. Within the latter policy, the Dutch government actively contributes to the establishment and operation of an interface between universities and the pharmaceutical industry, in order to improve the match between knowledge needs and knowledge supply (Nederlandse Vereniging van de Innoverende Farmacentische Industrie, 1995).

An intriguing final question is how can regional policy contribute to a strengthening of knowledge co-operation of the pharmaceutical industry with research institutes and small biotechnology firms? Given the need in the pharmaceutical industry for highly specialised expertise including biotechnology (often not available on a regional basis), it is not realistic to expect the emergence of close knowledge co-operation on a regional basis. However, an improved interface between the industry and research institutes, as well as new small biotechnology firms, may lead to gateways for regional companies to

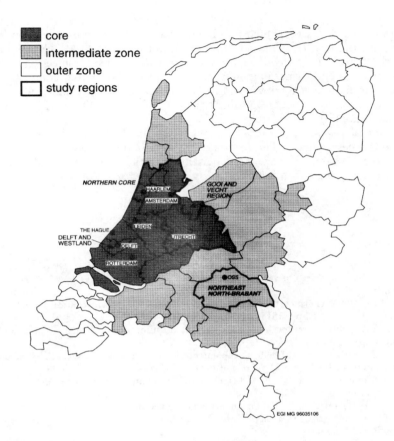

Figure 14.1
The Netherlands including the study regions
(Source: van Geenhuizen and van der Knaap, 1997)

global knowledge and for global companies to bring knowledge into the region. Furthermore, particular attention needs to be given to the creation of favourable conditions for the establishment of new biotechnology firms (e.g. spin-offs from knowledge institutes) and for their survival. Various needs of biotechnology start-ups can be satisfied on a local basis, such as the need for recruitment of personnel, cheap accommodation with potential for physical growth, and use of laboratory equipment (Feldman, 1985). Such policy may lead to the emergence of regional concentrations of biotechnology firms, such as in Leiden and Groningen. In this respect it would be useful to further advance the growth of regional (specialised) centres of expertise, with universities, academic hospitals, clusters of small biotechnology firms, and pharmaceutical companies as major regional knowledge creators. Such regional centres need also to provide excellent connections with global networks of expertise, so that regional actors can easily perform in global networks and global actors can connect with the regional expertise.

Connections are multidimensional in quality and type. They include electronic and library access to up-to-date research results in the world, and information about where research is being carried out and expertise is available (e.g. laboratories, institutes). They also include facilities to link with research databases in the world, and to participate or organise video-conferencing. Moreover, researchers need to be connected to the right sources which provide the latest information, and to know which networks are the best (i.e. know-who, in addition to know-how). Connections also include the availability of facilities for conferences and business-meetings, and facilities for informal and casual meetings. In terms of transportation connections mean easy access to high-speed railways and international airports. Finally, good connections mean a consistent image-building of centres of expertise throughout the outside world, based upon the particular specialisation.

Some initiatives have already been taken in the Netherlands, with a varying amount of success to date. The results of these initiatives need to be monitored, in order to take advantage of learning experiences, in the region involved and in other regions including high-tech industry.

REFERENCES

Amin, A. (1993) The globalisation of the economy: an erosion of regional networks?, in Grabher, G. (ed.) *The embedded firm. On the socioeconomics of industrial networks*, pp. 278–95, Routledge, London.
Anderson, M.J. (1993) Collaborative integration in the Canadian pharmaceutical industry, *Environment and Planning*, A, 25, pp. 1815–38.
Andersson, A.E. (1991) Creation, innovation and the diffusion of knowledge. General and specific economic impacts, *Sistemi Urbani*, 3, pp. 5–28.
Arora, A. and Gambardella, A. (1990) Complementarity and external linkages: the strategies of the large firms in biotechnology, *The Journal of Industrial Economics*, 38, pp. 361–79.
Boekema, F. and Kamman, D.J.F. (eds.) (1989) (in Dutch) *Socio-economic networks*, Wolters-Noordhoff, Groningen.
Camagni, R. (1991) Local 'milieu', uncertainty and innovation networks: towards a new dynamic theory of economic space, in Camagni, R. (ed.) *Innovation networks: spatial perspectives*, pp. 121–44, Belhaven Press, London.
Charles, D. and Howells, J. (1992) *Technology Transfer in Europe. Public and Private Networks*, Belhaven Press, London.
Coase, R. H. (1937) The nature of the firm, *Economica*, NS IV 13–16, pp. 386–405.
Daly, P. (1985) *The Biotechnology Business: A strategic analysis*, Frances Pinter, London.
Deeds, D.L. and Hill, C.W.L. (1996) Strategic alliances and the rate of new product development: an empirical study of entrepreneurial biotechnology firms, *Journal of Business Venturing*, 11, pp. 41–55.
Dosi, G., Pavitt, K. and Soete, L. (1990) *The Economics of Technical Change and International Trade*, New York University Press, New York.
Drewett, R., Knight, R. and Schubert, U. (1992) The Future of European Cities. The Role of Science and Technology, Report for the FAST Programme, Commission of the European Communities, Brussels.
Feldman, M.M.A. (1985) Biotechnology and local economic growth: the American pattern, in Hall, P. and Markusen, A. (eds.) *Silicon Landscapes*, pp. 65–79, Allen & Unwin, Boston.
Geenhuizen, M.S. van (1993) A Longitudinal Analysis of the Growth of Firms. The Case of the Netherlands, PhD Thesis, Erasmus University, Rotterdam.
Geenhuizen, M. van and Knaap, B. van der (1994) Dutch Textile Industry in a Global Economy, *Regional Studies*, 28, pp. 695–711.
Geenhuizen, M. van, Damman, M. and Nijkamp, P. (1996) The Local Environment as a Supportive Operator in Innovation Diffusion, Research Memorandum FEWEC, Free University, Amsterdam.
Geenhuizen, M. van and Knaap, B. van der (1997) Dynamics of R&D and regional networks in the Dutch pharmaceutical industry, *Journal of Social and Economic Geography*, (forthcoming).
Gertler, M. (1988) The limits to flexibility: comments on the post-Fordist vision of production and its geography, *Trans. Inst. Brit. Geogr.*, 17, pp. 259–78.

Hagedoorn, J. (1993) Strategic technology alliances and modus of cooperation in high-technology industries, in Grabher, G. (ed.) *The embedded firm. On the socioeconomics of industrial networks*, pp. 116–37, Routledge, London.
Håkansson, H. (1988) *Industrial Technological Development. A Network Approach*, Croom Helm, London.
Håkansson, H. and Johanson, J. (1993) The network as a governance structure, in Grabher, G. (ed.) *The embedded firm. On the socioeconomics of industrial networks*, pp. 35–51, Routledge, London.
Halliday, R.G., Drasdo, A.L., Lumley, C.E. and Walker, S.R. (1997) The allocation of resources for R&D in the world's leading pharmaceutical companies, *R&D Management*, 27, pp. 63–77.
Haug, P. (1995) Formation of biotechnology firms in the Greater Seattle region: an empirical investigation of entrepreneurial, financial, and educational perspectives, *Environment and Planning*, A, 27, pp. 249–67.
Howells, J. (1992) Pharmaceuticals and Europe 1992: the dynamics of industrial change, *Environment and Planning*, A, 24, pp. 33–48.
Kamann, D.J. (1993) Bottlenecks, Barriers and Networks of Actors, in Ratti, R. and Reichman, S. (eds.) *Theory and Practice of Transborder Cooperation*, pp. 65–101, Helbing & Lichtenhahn, Frankfurt am Main.
Kanter, R.M. (1995) *World Class: Thriving Locally in a Global Economy*, Simon and Schuster, New York.
Kohler, W. (1996) (in Dutch) Biotech increases opportunities for large pharmaceutical companies, *NRC-Handelsblad-Economie*, November 6.
Kramer, J.H.T. (1991) Formation as an economic-geographical concept, in Smidt, M. de and Wever, E. (eds.) *Complexes, formations and networks*, pp. 25–39, Royal Dutch Geographical Society, Utrecht.
Lambooy, J.G. (1991) Complexity, formations and networks, in Smidt, M. de and Wever, E. (eds.) *Complexes, formations and networks*, pp. 15–23, Royal Dutch Geographical Society, Utrecht.
Malerba, F. (1992) Learning by firms and incremental technical change, *The Economic Journal*, 102, pp. 845–59.
Ministry of Economic Affairs & Booz, Allen and Hamilton (1993) *Pharmaceutical R&D in the Netherlands – a strategic perspective*, Ministry of Economic Affairs, The Hague.
Ministry of Economic Affairs, Ministry of Education & Ministry of Agriculture (1995), (in Dutch) *Knowledge on the move*, Ministry of Economic Affairs, The Hague.
Nederlandse Vereniging van de Innoverende Farmacentische Industrie (various years) Jaarverslag, NEFARMA, Utrecht.
Organon (later AKZO-Pharma) (various years) Annual Report, Oss: Organon.
Porter, M.E. (1990) *The Competitive Advantage of Nations*, Macmillan, Basingstoke.
Powell, W.W. (1990) Neither market nor hierarchy: network forms of organisation, *Research in Organisational Behavior*, 12, pp. 295–336.
Richardson, G.B. (1972) The organisation of industry, *Economic Journal*, 82, pp. 883–96.
Sapienza, A.M. (1989) R&D collaboration as a global competitive tactic – Biotechnology and the ethical pharmaceutical industry, *R&D Management*, 19, pp. 285–95.
Saxenian, A.L. (1994) *Regional Advantage. Culture and Competition in Silicon Valley and Route 128*, Harvard University Press, Cambridge, MA.
Schoenberger, E. (1987) Technological and organisational change in automobile production: spatial implications, *Regional Studies*, 21, pp. 199–214.
Scott, A.J. (1988) *New industrial spaces: flexible production organisation and regional development in North America and Western Europe*, Pion, London.
Senker, J. (1996) National systems of innovation, organisational learning and industrial biotechnology, *Technovation*, 16, 5, pp. 219–29.
SER (Sociaal-Economisch Raad) (1995) (in Dutch) *Knowledge and Economy*, SER, Den Haag.
Storper, M. (1996) Innovation as Collective Action: Conventions, Products and Technologies, *Industrial and Corporate Change*, 5, 3, pp. 761–90.
Storper, M. & Scott, A.J. (1989) The geographical foundations and social regulation of flexible production complexes, in Wolch, J. & Dear, M. (eds.) The Power of Geography: How Territory Shapes Social Life, pp. 25–43, Unwin Hyman, Boston.
Storper, M. & Walker, R. (1989) The Capitalist Imperative: Territory, Technology and Industrial Growth, Basil Blackwell, New York.

Tausk, M. (1978) Organon, Dekker & Van de Vegt, Nijmegen.
Walsh, V. (1993) Demand, public markets and innovation in biotechnology, *Science and Public Policy*, 20, 3, pp. 138–56.
Wever, E. (1994) Industrieclusters: oude wijn in nieuwe zakken?, in Dijk, J. van and Florax, R. (eds.) (in Dutch) *Industry Policy, Regional Clusters and Markets*, pp. 35–46, Geopers, Groningen.
Williamson, O.E. (1985) *The Economic Institutions of Capitalism*, Free Press, New York.

CHAPTER 15

Inter-firm Links between Regionally Clustered High-Technology SMEs: A Comparison of Cambridge and Oxford Innovation Networks

CLIVE LAWSON, BARRY MOORE, DAVID KEEBLE,
HELEN LAWTON SMITH AND FRANK WILKINSON

INTRODUCTION

The 'resurgence of regional economies, and of territorial specialisation in an age of increasing ease in transportation and communication of inputs and outputs' is perhaps the major phenomenon in need of explanation in economic geography (Storper, 1995, p. 210). Furthermore, there has been a particularly strong clustering of firms in high-technology, information intensive sectors, sectors which one might have expected, given the enormous recent developments in the new information technologies, to be the least sensitive to the need for geographical proximity. In attempting to explain these developments, various recent contributions have focused upon the role and importance of various types of linkages or network of relations between firms. This paper is concerned with the nature and extent of such linkages and the extent to which they do indeed account for the observed clustering in 2 prominent high-technology regions in the UK, namely Cambridge and Oxford.

One notable effect of the burgeoning literature on the clustering of high-technology firms is that there has been a proliferation of new terminology, involving a whole range of different terms such as industrial districts, technological districts, technology districts, technological complexes, innovative milieux, nexuses of untraded interdependencies, and so on. Now whereas there are many clear differences in the theoretical positions that have given rise to these terms, the main concern here is with the similarities between them. Specifically, there has been a similar shift in focus in each of these positions as regards the kinds of linkage thought to be of particular importance. Brief reference to 3 such strands, and the recent change of focus within each, proves to be especially illustrative and serves to set the context for this study.

Some recent shifts in focus

The first strand is that arising out of a series of contributions under the heading of the 'Californian School'. Of interest here is the shift in focus from what has been termed *traded* to *untraded* interdependencies. In the early contributions of this school, the focus is upon the relationship between the division of labour, transactions costs and agglomeration. The (vertical) disintegration of production leads to increased transactions costs. Agglomeration then results from attempts to reduce these extra transactions costs (at least, those arising from geographical distance). However, the focus in such early contributions is predominately on (traded) input-output relations. More recently, Storper, drawing upon ideas from the technological trajectories literature (see Dosi and Orsenigo, 1985; Dosi, 1987; Arthur, 1989) and the technological learning literature (Lundvall, 1992; Lundval and Johnson, 1994), has argued that it is *untraded* interdependencies that explain the observed spatial patterns and that these 'cannot be easily accommodated within transactions-cost based theories' (1995, p.207). These untraded interdependencies cannot be captured by reference to input-output transactions or contract exchanges, but involve technological spillovers, various labour market characteristics, and conventions, rules and languages for developing, communicating and interpreting knowledge. A central point is that these untraded interdependencies not only take time to develop, but develop along certain trajectories (and are not 'time-reversible'). For example, in explaining why Silicon Valley shows no sign of weakening as an agglomeration, Storper explicitly argues that this is because 'geographically-constrained untraded interdependencies outlive geographically constrained input output linkages' (ibid p.209).

A similar shift in focus can be discerned in the literature on industrial districts inspired by the work of Marshall (1920, 1947). A firm's survival is taken to depend upon increased differentiation and more complex or sophisticated co-ordination – 'as a result of this greater subdivision, the parts of the system become increasingly mutually dependent and therefore necessarily co-operative' (You and Wilkinson, 1994, p.261). Two elements can be discerned in this move to increased interdependence which have clearly been brought out in the literature influenced by Marshall, especially work focusing upon the industrial districts of north east and central Italy. There is much concentration in the literature on transactions between firms in sequential stages in supply chains, frequent sharing of equipment, the possibility of jointly taking on larger orders, and upon large pools of appropriately skilled labour. Moreover, there is a special focus upon the particular forms of co-operation which take place in these districts, involving the sharing of technical information, subcontracting out to other (often less successful) competitors, and refraining from wage competition and labour poaching (Brusco and Sabel, 1981; Sabel and Zeitlin, 1985, pp.146–49; Lorenz, 1992). However, in explaining these linkages, or simply in elaborating them in more detail, 2 different emphases are evident. The first emphasis simply relates to the existence of external economies of scale which, although external to a particular firm, are internal to the productive system (e.g. industrial district) as a whole. The second emphasis relates to a general climate or 'industrial atmosphere' (see especially Bellandi, 1989; Becattini, 1990). In Marshall's work (1920, 1947) this aspect is mostly clearly brought out in his famous discussion of special and hereditary skills – where 'the mysteries of the trade become no mysteries; but are . . . in the air'. Here, however the emphasis is upon the network of conventions, rules, common understanding and so on, which make up the cultural, socio-economic 'industrial atmosphere' (Bellandi, 1989).

A third manifestation of this shift in focus is to be found within the GREMI[1] literature (Aydalot, 1986; Aydalot and Keeble, 1988; Camagni, 1991). In this body of work it is the local environment or *milieu* which is seen as the relevant unit of analysis, the focus being especially on the ability of the milieu to foster or facilitate *innovation*. In particular, the emphasis is upon the importance of a set, or a complex network, of mainly informal social relationships (Camagni, 1991). Innovations result from 'collective interactions' where a productive system, a technical culture and the various protagonists are linked (Crevoisier and Maillat, 1991). In attempting to distinguish the GREMI approach from others which emphasise the role of socio-cultural relationships, Camagni isolates what he terms static and dynamic approaches to the interpretation of economic space (both of which are aspects of the GREMI approach). Amongst the former, Camagni includes transaction costs and Marshallian ideas on external economies. When discussing the 'dynamic' aspects of the GREMI approach, Camagni points to the milieu as, on one hand, facilitating 'collective learning' and, on the other hand, reducing 'dynamic uncertainty'. The term 'collective learning', although not always used consistently within the GREMI literature, may be defined as the creation and further development of a base of common or shared knowledge among the individuals within a productive system. This allows both the co-ordination of action, and the resolution of problems (see Lazaric and Lorenz, 1996). Essentially, collective learning refers to that learning which is made possible through membership of some particular milieu (i.e. a set of local relationships making up a productive system). Shared knowledge here involves, for example, the establishment of a common language, technical know-how, and organisational conventions (see also Perrin, 1991).

In the GREMI literature the role of the milieu, as a mechanism for reducing uncertainty, similarly hinges on the relations and shared knowledge which facilitate acting capably. Various dimensions of uncertainty are distinguished and analysed within this approach (see especially Camagni, 1991). Uncertainty follows from the complexity of information (requiring a *search* function); from the problems of the *ex-ante* inspection of the qualitative features of inputs and equipment (requiring *screening*); and assuring customers of product quality (requiring *signalling*); from the problem of processing available information (involving *transcoding*); and of assessing the outcomes of one's own actions, and the actions of others. A distinction is then made between the types of linkages that serve to reduce uncertainty in each case. Collective information gathering and screening take place through the informal interchange of information between firms, while customer signalling is aided by milieu 'image', reputation and even co-operative advertising. Skilled labour mobility within the local labour market, customer-supplier technical and organisational interchange, imitation, application to local needs, or general purpose technologies and informal cafeteria effects, enable the transcoding function. A collective process of selecting decision routines results from managerial mobility, imitation, and co-operative decision-making through local associations or other bodies. Finally it is argued that an informal process of decision co-ordination is achieved via interpersonal linkages through families, clubs and associations, which has the advantage of easier and faster information circulation within similar cultural backgrounds. None of these factors is likely to be captured by any kind of input-output analysis, or study of simple (material) transactions.

Thus, to take stock, there have been clear common developments in all 3 of the approaches referred to above. Given an environment of organisational (vertical) disintegration, the links between smaller units increasingly have become the focus of attention. However, a concern with such linkages has increasingly been accompanied by a movement in focus away from simple input-output relations to a consideration of such factors as the less tangible absorptions and exchanges between agents and the qualitative nature of the rules, conventions, and social relations which allow agents to act in capable, innovative ways.

To point out that there are common underlying shifts in focus in these literatures should not be seen to imply any wide consensus on a substantive level. Various disputes and different emphases emerge once the nature and extent of such linkages is focused upon directly. For example, various criticisms exist to the effect that existing studies have over-emphasised the importance of local networks through focusing on the wrong sorts of data. Curran and Blackburn argue that whatever the observed spatial patterns, the 'local economy' is becoming less and less an integrated set of economic linkages, and so such linkages cannot be taken to explain any observed clustering. They also argue that these links have been growing weaker for some time, and that this has not been recognised because 'so much of the research on spatial aspects of the economy has been based on top-down, aggregate data analysis rather than on the closely observed activities of real people in real businesses in real places' (Curran and Blackburn, 1994, p.182). A different line of criticism is that the over concern with input-output thinking has led to a lack of attempts to document the actual nature of such relations in any serious way (Storper, 1995).

This paper is intended as a response to these types of criticism, while also considering the common shifts in focus noted above. In particular, its aim is to focus upon the nature and extent of local inter-firm linkages in 2 similar but different high-technology clusters. An attempt is made to focus upon more than simple input-output relations or external economies and to shed light on those factors conceptualised above as untraded interdependencies, industrial atmosphere or collective learning. In particular, some attempt is also made to highlight indicators of knowledge-based linkages which are vital to the innovativeness of high-technology clusters.

The rest of this paper is as follows. First a brief section is provided explaining the choice of regions, the sampling frame and the methodology used. The next 2 sections refer directly to results of a recent study carried out by the ESRC Centre for Business Research of Cambridge University in the Oxford and Cambridge regions[2]. There follows a brief concluding section.

SELECTION OF SAMPLES AND SURVEY METHODOLOGY

The CBR survey involved detailed face-to-face interviews and collection of data from 100 high-technology firms, 50 from each of the Cambridge and Oxford regions. The choice of Cambridge was an obvious one. Cambridgeshire recorded the largest volume of high-technology employment growth of all the UK counties between 1980 and 1990, with a further growth of 4,800 jobs or 17% between 1991 and 1995 (Keeble, 1989; 1994). By 1996, the county contained over 1,000 firms, overwhelmingly small and medium sized firms, based in high-technology sectors (Cambridgeshire County Council Research Group, 1996). Garnsey and Cannon-Brookes note that the apparent success of high-technology firms in the Cambridge region has been a major spur to all the main political parties in the process of establishing support for 'enterprise development' (Garnsey and Cannon-Brookes, 1993). Furthermore, they note that Cambridge is argued to be the country's 'undisputed centre for R&D' (Shirreff, 1991). However, it is fair to say that the early optimism about the prospects of Cambridge firms, especially as generated by the Segal Quince Wicksteed (SQW) report, *the Cambridge Phenomenon* (1985), has not been realised. In an attempt to explain this state of affairs, there has been much discussion of the nature of linkages within the region. For example, in the original SQW report, although the main phenomenon documented was the growth of small-scale, high-technology industry around the university, few direct links between the university and the region's firms were found to exist. The importance of the university's proximity was (and is) accepted, but the nature of its influence was unclear, and certainly not captured simply by direct transactions. Saxenian has argued that the early comparisons between Cambridge and Silicon Valley were simply misplaced (Saxenian, 1988). Although observers often cite social networks among local entrepreneurs as a sign of the region's growing potential, Saxenian found no evidence of such interaction – 'tenants of the Cambridge Science Park complain repeatedly that there is no interaction – social or technical – among firms there' (Saxenian, 1988, p.74).

The Oxford region shares some obvious similarities with that of Cambridge. Oxford houses one of the UK's premier universities and has 9 hospitals, 7 of which have university research departments. The region has several government research laboratories, many leading edge (mainly indigenous) small firms, and a number of major industrial research laboratories. 'Oxfordshire is an elite area . . . the site of private and public investment in innovation' (Lawton Smith, 1996). The area, in common with Cambridge, contains a large pool of skilled labour, a desirable environment for scientists and engineers to live, and a heavy specialisation in R&D. Moreover, neither region has benefited from the presence of major defence spending or other state policies (see Lawton Smith, 1990). However, many important differences exist between the Cambridge and Oxford regions, 2 of which are particularly significant here. First, there were significant differences in employment trends in the 2 regions during the 1980s as recorded by the 1981 and 1991 Censuses of Employment[3]. Whereas employment in high-technology firms and establishments in the Cambridge travel to work area grew by about 45% over this period, the Oxford travel to work area suffered a small high-technology employment decline of about 2%. Secondly, Oxfordshire has a reputation for traditional manufacturing, including a history of manufacturing-based linkages with the industrial Midlands, which is absent in the Cambridgeshire case.

Several definitions were considered in attempting to single out high-technology firms in these areas, including that of Buchart (1987). It was decided that no single definition was entirely suitable. Instead, existing data bases of high-technology firms were used which had been compiled on a case by case basis. For the Cambridge region, the County Council Research Group data base was utilised. For Oxford a similar but original data base was constructed from lists of firms compiled by Lawton Smith (1990), the Oxford Trust, and Oxfordshire County Council. Both these data bases were then refined and updated to provide a sampling frame of about 700 high-technology firms in Cambridge and 200 in Oxford. In each case the boundaries of the region concerned were drawn more tightly than the purely administrative county boundary so as to include only firms which were close enough to Cambridge or Oxford to have the possibility of easy daily functional contact and, less tangibly, to identify themselves as part of the Cambridge or Oxford milieu (for example, Huntingdon and Peterborough were excluded from the Cambridge population).

The sampling frame was then stratified in terms of the number of firms with over and under 100 employees and between services and manufacturing. A slightly higher number of larger firms were sampled given the likely disproportionately increased effects of these firms within the local economies. A division between services and manufacturing was made in proportion to the total numbers of such firms in each area. This provided a breakdown of firms as shown in Table 15.1. It should be stressed that virtually all the 'larger' firms surveyed employ less than 500 employees, about 95% of the final sample, and all but one in the Cambridge region, therefore being 'small or medium sized enterprises' (SMEs) using the EU traditional definition of less than 500 workers (Storey, 1994, p.6). It should also be pointed out that an examination of the high-technology firm size structure indicated a higher proportion of small firms in the Cambridge region, with about 74% of Cambridge firms employing less than 20 employees as compared to only 52% in the Oxford case (moreover only about 5% of Cambridge firms employ over 100 employees as compared with over 22% in the Oxford case). Response rates differed considerably for the 2 regions and are summarised in Table 15.1. The poorer Cambridge response almost certainly reflects over-surveying of firms in this region by many different research organisations.[4]

THE NATURE AND EXTENT OF LOCAL LINKAGES BETWEEN FIRMS

Information regarding the extent and nature of inter-firm links was gained via a series of related questions. In particular, firms were asked if they perceived themselves as

Table 15.1: *Achieved sample and response rates*

	Cambridge		Oxford	
Total contacted	130		75	
Achieved sample:				
Services < 100 employees	22	(42)[1]	17	(71)
Services > 100 employees	7	(65)	6	(54)
Manufacturing < 100 employees	16	(33)	22	(65)
Manufacturing > 100 employees	5	(29)	6	(55)
Total interviewed	50		50	
Response rate (%)	38		66	

[1] Response rate in each grouping

having close local links with other firms; what types of firms these links were with; how important these links are for their firms' development; and how important geographical proximity has been to the value of these links. Further information was gained by asking about the actual nature of the benefits and costs of such links. An important element of this questioning was that firms were continually asked about their perception of the importance of such links and what role proximity played in their usefulness.

Table 15.2: *Close local inter-firm links between small high-technology firms by sector (number and %)[1]*

Inter-firm links in	High-technology manufacturing	High-technology services	Total
Cambridge	17(34)	21(42)	38(76)
Oxford	14(28)	9(18)	23(46)
Oxford & Cambridge	31(31)	30(30)	61(61)
Number of firms in survey	48	52	100
of which: Cambridge	21	29	50
Oxford	27	23	50

Source for all the following tables:
ESRC Centre for Business Research Sample Survey
[1] % of total respondents

Table 15.2 shows the extent of inter-firm linkages between high-technology firms in the manufacturing and services sectors. These data, of course, exclude links with other local organisations such as universities or government research establishments. In total, 61% of the firms in the sample had close inter-firm links within the local economy. However inter-firm links are much more frequent in Cambridge where just over three quarters of the companies claimed to have close links with other local firms, compared with slightly less than half of the firms in Oxford. The most frequent local inter-firm linkages were to be found in the manufacturing sector of the Cambridge area, where 80% (17) of firms had links with other firms, while the least frequent instance was for firms in the Oxford service sector where only 39% (9) had close local inter-firm links.

The nature of local inter-firm links identified by Oxford and Cambridge firms as particularly important cover a wide spectrum. Inter-firm links based on sub-contracting were most frequent and accounted for 21% of all links (Table 15.3). Local customer/supplier links were also among the most important and together account for a further 36%. Eight (13%) firms were engaged in local collaborative activities of one kind or another, ranging from formal joint ventures, through the sharing of information and alliances for manufacturing, to collaboration on technical developments.

Table 15.3: *The nature of local inter-firm links established by high-technology companies (numbers and % of all firms with local links)*

Type of link	Number of firms	% of all inter-firm links
Sub-contracting	13	21
Suppliers	13	21
Customers	9	15
Collaboration	8	13
Consultants	4	7
Shareholders	3	5
Personal/historical	3	5
Subsidiary/satellite	2	3
Other	6	10
Total	61	100

Firms were asked to assess the importance of their links with other enterprises on a scale ranging from 1 to 5, where 1 equals completely unimportant and 5 equals highly important. Table 15.4 shows the frequency of firms ranking their links with a score of 4 or 5 for each of 6 possible types of link. The majority of high-technology firms experiencing close links with other local firms regarded such links as important, or very important, and this was particularly so in the Oxford case (i.e. 87%). Links with local suppliers and subcontractors were cited most frequently as important in both Cambridge and Oxford. About one third of firms perceived links with local service providers as very important. Research collaboration most differentiates Cambridge and Oxford in terms of the importance of different types of links with 39% of high-technology firms claiming them to be important in the Oxford local economy and only 11% in Cambridge.

Table 15.4: *The importance of different inter-firm links (number and % of firms with links)*[1]

Type of link	Cambridge	Oxford	Total
Customers	8(21)	9(39)	17(28)
Suppliers/subcontractors	17(45)	13(57)	30(49)
Firms providing services	12(32)	9(39)	21(34)
Research collaborators	4(11)	9(39)	13(21)
Firms in your line of business	4(11)	3(13)	7(11)
Others	1(3)	0(0)	1(2)
Total	25(66)	20(87)	45(74)

[1] % ranking 4 or 5 of all firms with links in each local area

Whereas the evidence above demonstrates that local links are both frequent and of importance to high-technology firms in the Cambridge and Oxford clusters, the question remains as to how important geographical proximity is to the value of such links. In other words, although links are clearly valuable, how much of this value actually depends upon geographical proximity? As Table 15.5 shows, more than two thirds of firms regarded geographical proximity as very important in the development and maintenance of local inter-firm links.

Table 15.5: *The importance of geographical proximity in the development of local inter-firm links (number and % of firms with links)*[1]

Type of link	Cambridge	Oxford	Total
Customers	6(16)	5(22)	11(18)
Suppliers/subcontractors	15(39)	12(52)	27(44)
Firms providing services	10(26)	11(48)	21(34)
Research collaborators	5(13)	5(22)	10(16)
Firms in your line of business	3(8)	2(9)	5(8)
Others	1(3)	0(0)	1(2)
Total	23(61)	19(83)	42(69)

[1] % ranking 4 or 5 of all firms with links in each local area

A greater proportion (83%) of linked Oxford high-technology firms than Cambridge linked firms (i.e. 61%) regarded geographical proximity as important to such links. Interestingly, although links with customers were crucial overall (see Table 15.6), geographical proximity was not a vital aspect of these links – especially in relation to links with suppliers, subcontractors and firms providing services.

Table 15.6: *The importance of links outside the region and their increased usefulness if they were inside the region (number and % of firms with links)*[1]

Type of Link	Cambridge		Oxford		Total	
	(a)[2]	(b)[3]	(a)	(b)	(a)	(b)
Customers	32(84)	10(26)	18(78)	9(39)	50(82)	19(31)
Suppliers/subcontractors	17(45)	17(45)	14(61)	17(74)	31(51)	34(56)
Firms providing services	4(11)	10(26)	8(35)	14(61)	12(20)	24(39)
Research collaborators	9(24)	10(26)	11(48)	8(35)	20(33)	18(30)
Firms in your line of business	7(18)	4(11)	6(26)	5(22)	13(21)	9(15)
Others	0(0)	0(0)	0(0)	0(0)	0(0)	0(0)
Total	34(89)	23(61)	21(91)	19(83)	55(90)	42(69)

[1] % ranking 4 or 5 of all firms with links in each local area
[2] Importance of links outside the region
[3] Increased usefulness if links inside the region

Geographical proximity is clearly important for supplier and service provision linkages and much less important for customers. This almost certainly reflects the very high export intensity of the high-technology SMEs surveyed (Keeble et al., 1997), with no less than 45% of firms reporting exports as accounting for at least 40% sales, let alone sales elsewhere in the UK. One sixth of companies regarded proximity as important for research collaboration which was slightly less than the proportion claiming important local links for research (i.e. 21%, see Table 15.5). To explore further the importance that high-technology firms attach to geographical proximity, they were asked how important their links were with firms outside the Cambridge/Oxford region and whether such links would be more useful if the firms were more closely located inside the local area. The majority of important customer links were outside the region for high-technology firms in both Cambridge and Oxford at 84% and 78% respectively. In Cambridge just over one quarter of firms (with inter-firm links) said that links with

INTER-FIRM LINKS BETWEEN REGIONALLY CLUSTERED HIGH-TECHNOLOGY SMEs 189

customer firms would be more useful if they were inside the region, whereas in Oxford the equivalent proportion was somewhat higher at 39%.

For suppliers and subcontractors, important links outside the region were less frequent than in the case of customers, but a much greater proportion of firms thought that such links would be more useful if their suppliers/subcontractors were located inside the local area. Where inter-firm links outside the region exist for the purpose of collaborative research, 30% of firms indicated that such links would be more useful if the collaborating firms were both inside the local area.

The evidence in Tables 15.5 and 15.6 above not only reveals the importance of geographical clustering for inter-firm links, but also highlights its particular importance for supplier/subcontractor relationships. Much more limited benefits from clustering, however, are perceived for other types of inter-firm links. The interesting question now is, what are the nature of the benefits that flow from inter-firm links and which of these benefits are more easily gained by which firms within their respective regions?

Given the importance attached to supplier links by high-technology firms, it is perhaps not surprising that assurance of quality and timely delivery of supplies were cited by 56% and 49% of firms respectively as important benefits of close inter-firm links. Moreover similar proportions of firms (44% and 49%) thought that such benefits could be more easily secured from inter-firm links within the local area. A smaller proportion of firms in Cambridge than in Oxford perceived proximity as facilitating the acquisition of these benefits and this, no doubt, partly reflects the higher proportion of Oxford firms linking closely with their suppliers (Table 15.4). Other benefits flowing from inter-firm high-technology links relate to new products and responsiveness to market requirements. About half the high-technology firms in both Oxford and Cambridge that had formed inter-firm links saw improvements in the amount and

Table 15.7: *The benefits from inter-firm links and the importance of proximity in securing these benefits (number and %)*[1]

Main benefits of links	Cambridge		Oxford		Total	
	(a)[2]	(b)[3]	(a)	(b)	(a)	(b)
Improving amount of information about new products	20(53)	12(32)	10(43)	11(48)	30(49)	23(38)
Improving the quality of information about new products	20(53)	13(34)	13(57)	11(48)	33(54)	24(39)
Improving access to research findings	9(24)	10(26)	10(43)	10(43)	19(31)	20(33)
Assuring a satisfactory quality of supplies	19(50)	14(37)	15(65)	13(57)	34(56)	27(44)
Assuring a timely delivery of supplies	15(39)	14(37)	15(65)	16(70)	30(49)	30(49)
Greater responsiveness to market requirements	20(53)	6(16)	14(61)	8(35)	34(56)	14(23)
More effective or innovative R&D	18(47)	12(32)	10(43)	11(48)	28(46)	23(38)
Other	2(5)	2(5)	0(0)	2(9)	2(3)	4(7)
Total	34(89)	25(66)	20(87)	20(87)	54(89)	45(74)

[1] % of all firms with links in each local area
[2] Main benefits from inter-firm links
[3] Benefits more easily gained from links inside the region

quality of information as the main benefits from such links. Moreover about one third of firms thought that such benefits could be more easily gained from links with firms within the local economy.

In Table 15.7 a similar picture emerges with respect to the benefits associated with improved market responsiveness, although a smaller overall proportion of firms believe that proximity facilitates securing these benefits. Other main benefits from inter-firm links relate to more effective or innovative R&D and this is perceived to be important for 47% of Cambridge firms and 43% of Oxford firms. About one third of Cambridge firms and half of Oxford firms thought that such benefits could be more easily secured with geographical proximity. Finally, improved access to research findings was seen as a main benefit from inter-firm links by just under one quarter of Cambridge firms and 43% of Oxford firms, but only one third of firms from both regions thought that proximity facilitated the securing of these benefits.

Table 15.8: *The main risks of establishing close links with other firms (number and %)*[1]

Main risks of links	Cambridge		Oxford		Total	
	(a)[2]	(b)[3]	(a)	(b)	(a)	(b)
Loss of property rights in new products/processes	6(16)	4(11)	5(22)	6(26)	11(18)	10(16)
Loss of control of confidential business information	12(32)	6(16)	7(30)	5(22)	19(31)	11(18)
Over-reliance/dependence on specific firms	17(45)	4(11)	16(70)	5(22)	33(54)	9(15)
Other	1(3)	0(0)	0(0)	1(4)	1(2)	1(2)
Total	23(61)	8(21)	18(78)	9(39)	41(67)	17(28)

[1] % of firms with local links
[2] Main risks of inter-firm links
[3] Risks reduced by links

Local inter-firm links not only generate benefits for the firms involved but also involve potential costs. Table 15.8 shows that the most frequently cited overall risk associated with local inter-firm links (i.e. 54%) is over-reliance or dependence on specific firms, and this was particularly the case in Oxford where 70% of firms saw this as a major risk. Loss of control of confidential business information was a main risk for about 30% of firms and loss of property rights for about one fifth of firms.

Close inter-firm links may provide the circumstances and relationships of trust where risks may be reduced and overall 28% of firms believed this to be the case (column (b) Table 15.8). In this respect there was a significant difference between Oxford and Cambridge with 39% of firms in the former perceiving a reduction in risk compared with only 21% in the latter.

Interestingly, when asked how else risks might be reduced, high-technology firms in Oxford emphasised personal relationships and experience of fair trading, whereas firms in Cambridge were more inclined to a legalistic approach (Table 15.9).

INDICATORS OF A KNOWLEDGE-BASED MILIEU

Much of the recent literature referred to at the beginning of this paper stressed the importance of the role of links, especially within high-technology clusters, in providing necessary information for innovative activities. To assess this claim, information was

2. This research was funded by the UK Economic and Social Research Council, whose essential financial support is gratefully acknowledged.
3. Unpublished Census of Employment figures were obtained from NOMIS (National Online Manpower Information Service) in Durham. High-technology sectors were defined according to Butchart, 1987.
4. Piloting of the initial questionnaire took place with the help of the owner-managers of several local firms, the Director of St John's Innovation Centre, Cambridge, and a 'user workshop' made up of representatives from various local Cambridge firms and other interested bodies. Data collection took the form of face to face interviews between late November 1995 and May 1996. Each interview took between 45 minutes and 4 hours, with a mean length close to 1½ hours. All interviews were carried out by members of the research team. We should also like to acknowledge our sincere gratitude for the help given us by the owner-managers and directors of the surveyed firms, without which this study would not have been possible.
5. It is worth noting that this data only concerns ongoing linkages between firms. A vital local source of innovating activity in the 2 regions is the high level of new firm spin-off and start-up activity, based on the exploitation of ideas originating in other (local) firms or institutions. This is documented below.

REFERENCES

Arthur, W.B. (1989 Competing technologies, increasing returns and lock-in by historical events, *The Economic Journal*, 99, pp. 116–31.
Aydalot, P. (ed.) (1986) *Milieux Innovators en Europe*, GREMI, Paris.
Aydalot, P. and Keeble, D. (eds.) (1988) *High Technology Industry and Innovative Environments: the European Experience*, Routledge, London.
Becattini, G. (1990) The Marshallian Industrial District as a Socio-economic Notion, in Pyke et al., *Industrial districts and inter-firm co-operation in Italy*, International Institute for Labour Studies, Geneva.
Bellandi, M. (1989) The industrial district in Marshall, in Goodman et al. (eds.) *Small Firms and Industrial Districts in Italy*, Routledge, London.
Butchart, R.L. (1987) A new UK definition of the high technology industries, *Economic Trends*, 400, pp. 82–8.
Brusco, S. and Sabel, C. (1981) Artisan production and economic growth, in Wilkinson, F. (ed.) *Dynamics of labour market segmentation*, Academic Press, London.
Camagni, R. (ed.) (1991) *Innovation Networks: Spatial Perspectives*, Belhaven Press, London.
Cambridge County Council Research Group (1996) *The Hi-Technology 'Community' in Cambridgeshire (end 1995)*, Cambridgeshire County Council, Cambridge.
Crevoisier, O. and Maillat, D. (1991) Milieu, industrial organisation and territorial production systems: towards a new theory of spatial development, in Camagni, R. (ed.) *Innovation Networks: Spatial Perspectives*, Belhaven Press, London.
Curran, J. and Blackburn, R. (1994) *Small Firms and Local Economic Networks: The Death of the Local Economy?*, Paul Chapman, London.
Dosi, G. (1987) Institutions and markets in a dynamic world, SPRU Discussion Paper, No. 32, Brighton.
Dosi, G. and Orsenigo, L. (1985) Order and change: an exploration of markets, institutions and technology in industrial dynamics, SPRU Discussion Paper, No. 22, Brighton.
Garnsey, E. and Cannon-Brookes, A (1993) The Cambridge phenomenon revisited; aggregate change among Cambridge high technology companies since 1989, *Entrepreneurship and Regional Development*, 5, 1, pp. 179–207.
Keeble, D. (1989) High-Technology Industry and Regional Development in Britain: The Case of the Cambridge Phenomenon, *Environment and Planning C, Government and Policy*, 7, 2, pp. 153–72.
Keeble, D. (1994) Regional influences and policy in new technology-based firm creation and growth, in Oakey, R. (ed.) *New Technology-Based Firms in the 1990s: Vol I*, Paul Chapman, London.
Keeble, D., Lawson, C., Lawton Smith, H., Moore, B. and Wilkinson, F. (1997) Internationalisation Processes, Networking and Local Embeddedness in Technology-Intensive Small Firms, ESRC Centre for Business Research, Working Paper No. 53, Cambridge University.

Lawton Smith, H. (1990) The Location of Innovative Industry: the Case of Advanced Technology Industry in Oxfordshire, Research Paper 44, School of Geography, Oxford.

Lawton Smith, H. (1996) Local Links: The Oxfordshire Experience, Seminar Paper on barriers to technology transfer, Brunel University.

Lazaric, N. and Lorenz, E.H. (1996) Trust and Organisational Learning During Inter-firm Co-operation, in Lazaric, N. and Lorenz, E.H. (eds.) *The Economics of Trust and Learning*, Edward Elgar, London.

Lorenz, E.H. (1992) Trust, community and co-operation: towards a theory of industrial districts, in Storper, M. and Scott, A.J. (eds.) *Pathways to Industrialisation and Regional Development*, Routledge, London.

Lundvall, B.A. (1992) *National Systems of Innovation: Toward a theory of Innovation and Interactive Learning*, Frances Pinter, London.

Lundvall, B.A. and Johnson, B. (1994) The learning economy, *Journal of Industrial Studies*, 1, 2, pp. 23–42.

Marshall, A. (1920) *Industry and Trade*, Macmillan, London.

Marshall, A. (1947) *Principles of Economics*, Macmillan, London.

networks and milieux, in Camagni, R. (ed.) *Innovation Networks: spatial perspectives*, Belhaven Press, London.

Sabel, C. and Zeitlin, J. (1985) Historical alternatives to mass production: politics, markets and technology in nineteenth-century industrialisation, *Past and Present*, 108, pp. 133–76.

Saxenian, A (1988) The Cheshire Cat's grin: innovation and regional development in England, *Technology Review*, Feb/March, p. 67.

Segal Quince Wicksteed (1985) *The Cambridge Phenomenon: The Growth of High Technology Industry in a University Town*, SQW Limited, Cambridge.

Shirreff, D. (1991) The Business Guide, *Cambridge Business Magazine*, February, pp. 89–101.

Storey, D.J. (1994) *Understanding the Small Business Sector*, Routledge, London.

Storper, M. (1995) The resurgence of regional economies, ten years later: the region as a nexus of untraded interdependencies, *European Urban and Regional Studies*, 2, 3, pp. 191–221.

You, J.I. and Wilkinson, F. (1994) Competition and co-operation: toward understanding industrial districts, *Review of Political Economy*, 6, 3.